ENGINEERING APPLICATIONS OF CORRELATION AND SPECTRAL ANALYSIS

ENGINEERING APPLICATIONS OF CORRELATION AND SPECTRAL ANALYSIS

JULIUS S. BENDAT

Mathematical Consultant in Random Data Analysis

ALLAN G. PIERSOL

Senior Scientist, Bolt Beranek & Newman, Inc.

A Wiley-Interscience Publication

JOHN WILEY & SONS

New York · Chichester · Brisbane · Toronto

Library of Congress Cataloging in Publication Data

Bendat, Julius S
   Engineering applications of correlation and spectral
analysis.

   "A Wiley-Interscience publication."
   Bibliography: p.
   Includes index.
   1. Engineering—Statistical methods. 2. Correlation
(Statistics) 3. Spectral theory (Mathematics)
I. Piersol, Allan G., joint author. II. Title.

TA340.B427   519.5'37'02462   79-25926
ISBN 0-471-05887-4

Printed in the United States of America

10 9 8 7 6 5 4 3 2 1

PREFACE

Since the writing of our previous 1971 book (*Random Data: Analysis and Measurement Procedures*, Wiley-Interscience), there has been a rapid expansion in the engineering applications of correlation and spectral analysis techniques. This is primarily due to the growing availability of relatively inexpensive computer equipment that can rapidly execute the necessary computations, accompanied by new ideas on how to model and interpret computed results to solve many previously difficult problems. This text supplements the theory and data processing procedures of the 1971 edition by discussing practical matters and engineering applications of correlation and spectral functions, and should not be considered as a replacement or revision of the earlier book. Results contained here are largely based on various industrial and government projects in which we have been engaged since 1971, and preparation of this book was enhanced by lectures delivered by us on these subjects at short courses held in the United States and Europe.

Because of the nature of our work, the illustrations presented in this text are primarily concerned with applications to acoustics, mechanical vibrations, system identification, and fluid dynamics problems in the aerospace, automotive, industrial noise control, civil engineering, and oceanographic fields. However, the techniques and principles developed herein have wide applications to many other engineering fields, as well as to biomedical and economic time series analysis.

Our main objectives in writing this book were to answer the following important questions: Assuming one has the required hardware and software to compute correlation and spectral estimates for various specific engineering applications of interest, (1) What data should be collected? (2) What practical problem areas exist and how can they be handled? (3) What particular functions should be computed? (4) How should data be processed to reduce statistical bias and random errors? (5) How should computed quantities be interpreted so as to give physically meaningful results? Answers to these questions are possible only from experience, where solutions require a knowledge of the engineering "art" in dealing with actual data and actual systems, as well as a knowledge of the mathematical "theory" to predict results for ideal data and ideal systems. This book attempts to bridge the differences between measured results from engineering practice as opposed to expected theoretical results from analytical models by discussing the details from many engineering examples.

Measured results from correlation functions are compared to measured results from spectral density, coherence, and phase functions to bring out significant features in each type of analysis. These comparisons demonstrate for various engineering applications where it is appropriate to conduct one type of analysis over another and how supporting information can be obtained. The critical role played by good engineering judgment is emphasized in the selection of record lengths and digital data processing parameters to minimize statistical bias errors and random errors. Attention to these matters can greatly improve the interpretation and confidence in accepting conclusions drawn from such measured results.

It is assumed that the reader has a working knowledge of calculus, Fourier series, and complex variable theory. It is also assumed that the reader is familiar with system response characteristics and the basic concepts of probability and statistics, but for completeness, these matters are briefly reviewed in the first two chapters. Basic principles of correlation and spectral density analysis are presented in Chapter 3. Traditional procedures for analyzing single input/single output relationships and for estimating system properties and responses are detailed in Chapters 4 and 5. This material includes discussions of ordinary coherence functions, coherent output spectra, the effects of input and output measurement noise and feedback, the use of external excitation, and the estimation of frequency response functions.

Chapters 6 and 7 are concerned with problems related to time delay and phase lag estimates. Chapter 6 details the practical considerations involved in the identification of multiple propagation paths and velocities for both dispersive and nondispersive media based on input/output measurements or output measurements only. Chapter 7 addresses the related single input/multiple output problem where output measurements alone are available for analysis. Included are applications to the location of energy sources and the estimation of system response properties.

Chapters 8, 9, and 10 are concerned with the analysis of multiple input/output problems and applications of multiple and partial coherence functions. Fundamental ideas are explained in Chapter 8 which are valid for both single and multiple output problems. Chapter 9 deals with the important subject of source identification in multiple source problems for both physically correlated and uncorrelated cases. Such practical considerations as measurement interference and reverberation effects are detailed with illustrations. Computational algorithms for efficient digital data processing operations are derived and discussed in Chapter 10. Relationships are explained fully for two input systems prior to treating the general case of multiple inputs.

Chapter 10 also develops procedures for laboratory simulation of spectral density matrices representing autospectra and cross-spectra that can exist among an arbitrary number of multiple records.

The final Chapter 11 contains the latest available practical statistical error analysis formulas for computing spectral density functions, coherence functions, frequency response functions, and other related functions required in analyzing single input/output and multiple input/output problems. Statistical errors in computing probability functions and correlation functions are summarized in Chapters 2 and 3. These formulas are simple to apply and indicate how much data should be collected to achieve specified experimental results, as well as how to evaluate and interpret present and past results.

Although this book was written primarily for practicing engineers and scientists, it lends itself to seminar-type courses where professors can give lectures on each chapter and students can then study further special topics. Measurements should be carried out, if possible, using whatever equipment and computer programs are available at the schools to illustrate similar experimental results to those discussed in the book. This will help students appreciate the engineering problems that can occur, the importance of satisfying assumptions in models under consideration, and the significance of various statistical error analysis formulas to evaluate computed functions. The book will thus be a definitive guide to graduate students for advanced degrees doing experimental work involving correlation and/or spectral analysis of data, as well as to the professors for their own research projects.

We wish to thank friends all over the world who have supported this work by arranging for our participation in their engineering applications, and by sponsoring and attending our short courses. Special appreciation goes to our secretary, Connie Miller, for her dedication and assistance in producing this book, and to Ingrid Salazar for her contributions to the artwork.

JULIUS S. BENDAT
ALLAN G. PIERSOL

Los Angeles, California
January 1980

CONTENTS

GLOSSARY OF SYMBOLS

ENGINEERING APPLICATIONS OF CORRELATION AND SPECTRAL ANALYSIS

CHAPTER 1

INTRODUCTION AND BACKGROUND

Physical phenomena of common interest in engineering are usually measured in terms of an amplitude versus time function, referred to as a *time history record*. The instantaneous amplitude of the record may represent any physical quantity of interest, for example, displacement, velocity, acceleration, pressure, angle, temperature, and so on. Similarly, the time scale of the record may represent any appropriate independent variable, for example, relative time, spatial location, angular position, and so on. There are certain types of physical phenomena where specific time history records of future measurements can be predicted with reasonable accuracy based on one's knowledge of physics and/or prior observations of experimental results, for example, the force generated by an unbalanced rotating wheel, the position of a satellite in orbit about the earth, and the response of a structure to a step load. Such phenomena are referred to as *deterministic*, and methods for analyzing their time history records are well known. Many physical phenomena of engineering interest, however, are not deterministic, that is, each experiment produces a unique time history record which is not likely to be repeated and cannot be accurately predicted in detail. Such data and the physical phenomena they represent are called *random*.

This book is concerned with the general interpretations and applications of random data analysis with emphasis on the use of correlation and spectral density functions. As an introduction to this subject, a brief review is presented here of the characteristics of random data, the fundamentals of Fourier series and transform analysis, and the basic response properties

1

of physical systems. More detailed developments of this background material are available from the references.

1.1 CHARACTERISTICS OF RANDOM DATA

As just mentioned, a physical phenomenon and the data representing it are considered random when a future time history record from an experiment cannot be predicted within reasonable experimental error. In such cases, the resulting time history from a given experiment represents only one physical realization of what might have occurred. To fully understand the data, one should conceptually think in terms of all the time history records that could have occurred, as illustrated in Figure 1.1. This collection of all time history records $x_i(t)$, $i = 1, 2, 3, \ldots$, which might have been produced by the experiment, is called the *ensemble* that defines a *random process* $\{x(t)\}$ describing the phenomenon.

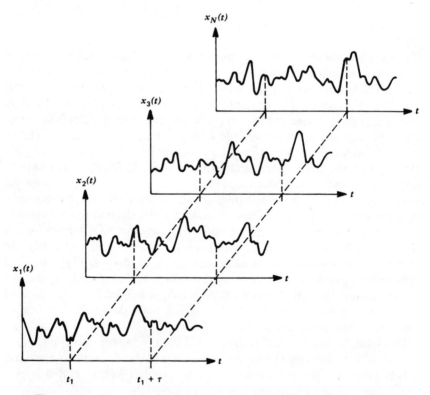

Figure 1.1 Ensemble of time history records defining a random process.

1.1.1   *Stationary Data*

Given an ensemble of time history records $\{x(t)\}$ describing a phenomenon of interest, the average properties of the data can be readily computed at any specific time t_1 by averaging over the ensemble. For example, the average value and the average squared value at time t_1, called the *mean value* and the *mean square value* of the data, respectively (to be developed more thoroughly in Chapter 2), are given by

$$\mu_x(t_1) = \lim_{N \to \infty} \frac{1}{N} \sum_{i=1}^{N} x_i(t_1) \tag{1.1}$$

$$\psi_x^2(t_1) = \lim_{N \to \infty} \frac{1}{N} \sum_{i=1}^{N} x_i^2(t_1) \tag{1.2}$$

Furthermore, the average product of the data values at times t_1 and $t_1 + \tau$, called the *autocorrelation* at time delay τ (to be developed more thoroughly in Chapter 3), is given by

$$R_{xx}(t_1, \tau) = \lim_{N \to \infty} \frac{1}{N} \sum_{i=1}^{N} x_i(t_1) x_i(t_1 + \tau) \tag{1.3}$$

An endless number of other higher order average values can be computed in a similar way. For the general case where one or more of these average values of interest varies with changes in the time t_1, the data are said to be *nonstationary*. For the special case where all average values of interest remain constant with changes in the time t_1, the data are said to be *stationary*. For stationary data, the average values at all times can be computed from appropriate ensemble averages at a single time t_1.

1.1.2   *Ergodic Data*

For almost all stationary data, the average values computed over the ensemble at time t_1 will equal the corresponding average values computed over time from a single time history record. For example, the average values in Equations 1.1 through 1.3 may be computed in most cases by

$$\mu_x = \lim_{T \to \infty} \frac{1}{T} \int_0^T x(t)\, dt \tag{1.4}$$

$$\psi_x^2 = \lim_{T \to \infty} \frac{1}{T} \int_0^T x^2(t)\, dt \tag{1.5}$$

$$R_{xx}(\tau) = \lim_{T \to \infty} \frac{1}{T} \int_0^T x(t)x(t + \tau)\, dt \tag{1.6}$$

where $x(t)$ is any arbitrary record from the ensemble $\{x(t)\}$. The justification for the above results comes from the *ergodic theorem* [1.1, 1.2], which states that, for stationary data, the properties computed from time averages over individual records of the ensemble will be the same from one record to the next and will equal the corresponding properties computed from an ensemble average over the records at any time t if

$$\frac{1}{T} \int_{-T}^{T} |R_{xx}(\tau) - \mu_x^2|\, d\tau \to 0 \qquad \text{as } T \to \infty \tag{1.7}$$

In practice, violations of Equation 1.7 are usually associated with the presence of periodic components in the data. Equation 1.7 is a sufficient but not a necessary condition for ergodicity, which means that time averages are often justified even when periodic components are present or other violations of Equation 1.7 occur; one simply must be more careful in such cases to confirm that the time averaged properties of different records are the same.

1.1.3   *Statistical Sampling Considerations*

The number of time history records that might be available for analysis by ensemble averaging procedures, or the length of a given sample record available for analysis by time averaging procedures, will always be finite; that is, the limiting operations $N \to \infty$ in Equations 1.1 through 1.3 and $T \to \infty$ in Equations 1.4 through 1.6 can never be realized in practice. It follows that the average values of the data can only be estimated and never computed exactly. The resulting error in the computation of an average value due to these finite sampling considerations is of major importance to the interpretations and applications of the analyzed data. Hence considerable attention is given in this book to the definition of statistical sampling errors for those data properties of interest for common applications. Important error formulas are discussed in Chapter 11 and in Sections 2.4 and 3.4.

It should be emphasized here that the errors presented in this book are only the statistical sampling errors inherent in analysis operations on finite amounts of data. Beyond these sampling errors are a multitude of other potential errors that may have accumulated during the data acquisition and preprocessing operations. Included are the errors that might arise from the measurement transducers, signal conditioning equipment, magnetic tape recording and/or telemetry if used, analog-to-digital conversion, and pre-analysis conditioning operations. All of these potential sources of error demand careful attention and must be diligently controlled or properly accounted for by accurate calibration procedures. It is assumed in this book that all such errors have been controlled or corrected, and that the time history records received for analysis accurately represent the physical phenomenon of interest. Further discussions of data acquisition and pre-processing errors are presented in Chapter 7 of Reference 1.3.

1.1.4 *Other Practical Considerations*

Every effort is usually made in practice to design experiments that will produce stationary data because the necessary analysis procedures for nonstationary data are substantially more difficult. In most laboratory experiments, one can usually force the results of the experiment to be stationary by simply maintaining constant experimental conditions. For example, if one is interested in the surface pressures inside a pipe due to high velocity air flow, stationary data will be generated if the air flow velocity, density, and temperature are held constant during each experiment. Typical time history records from such an experiment repeated under identical conditions are shown in Figure 1.2. It is clear from visual inspection that these data are stationary and ergodic.

In many field experiments as well, there is no difficulty in performing the experiments under constant conditions to obtain stationary data. There are, however, some exceptions. One class of exceptions is when the nature of the experiment requires the mechanisms producing the data of interest to be time dependent. Examples include the ground acoustic pressures due to an aircraft fly over and the vibration of a spacecraft structure during launch. In these cases, the experiments can hypothetically be repeated to obtain an

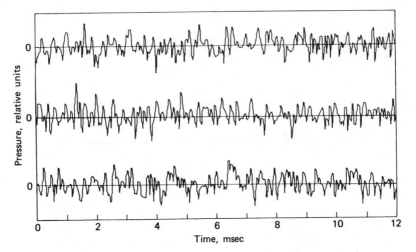

Figure 1.2   Ensemble of pipe-flow-pressure time history records.

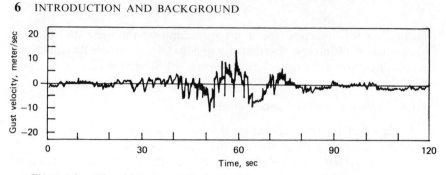

Figure 1.3 Time history record of nonstationary atmospheric gust velocities.

ensemble of records that properly represent the nonstationary phenomenon of concern. A second and more difficult class of exceptions is where the basic parameters of the mechanisms producing the data are acts of nature that cannot be controlled by the experimenter. Examples include time history data for ocean wave heights and atmospheric gust velocities. An illustration of nonstationary gust velocity data is shown in Figure 1.3. In these cases,

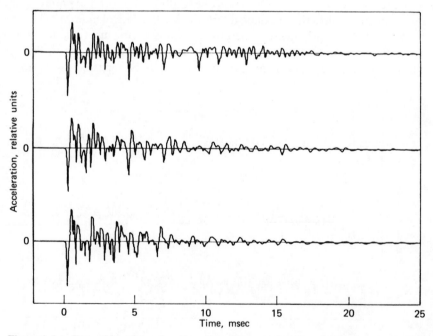

Figure 1.4 Ensemble of acceleration time history records of transient structural impacts.

one cannot even design repeated experiments that would produce a meaningful ensemble. You simply take what you get. The usual approach in analyzing such data is to select from the available records quasistationary segments that are sufficiently long to provide statistically meaningful results for the existing conditions. In some cases, special nonstationary models can be employed that permit the data to be decomposed into stationary random and nonstationary deterministic components for analysis. Such methods and other techniques for analyzing nonstationary data are discussed in Chapter 10 of Reference 1.3.

This book is concerned primarily with the interpretations and applications of stationary data analysis. However, one type of nonstationary data is considered, specifically, *transient* data that result from short-duration nonstationary phenomena with a clearly defined beginning and end. An illustration is presented in Figure 1.4 that shows an ensemble of acceleration time history records resulting from a structural impact load. Data of this type can be analyzed and interpreted by procedures that are closely related to the techniques appropriate for stationary data, as will be noted in Section 4.1.2.

1.2   FOURIER SERIES AND TRANSFORMS

Fourier series expansions and transformations are fundamental to the analysis techniques and applications developed and illustrated in this book. Hence a brief review is given here of some basic Fourier series and Fourier transform relationships. More detailed discussions are available from the many textbooks on this subject.

1.2.1   *Fourier Series*

Consider any periodic record $x(t)$ of period T. Then for any value of t

$$x(t) = x(t \pm kT) \qquad k = 1, 2, 3, \ldots \tag{1.8}$$

The fundamental frequency f_1 satisfies

$$f_1 = \frac{1}{T} \tag{1.9}$$

With few exceptions, such periodic data can be expanded in a *Fourier series* according to the following formula:

$$x(t) = \frac{a_0}{2} + \sum_{k=1}^{\infty} (a_k \cos 2\pi f_k t + b_k \sin 2\pi f_k t) \tag{1.10}$$

where

$$f_k = kf_1 = \frac{k}{T} \qquad k = 1, 2, 3, \ldots$$

Thus $x(t)$ is described in terms of sine and cosine waves at discrete frequencies spaced $\Delta f = f_1$ apart. The coefficients $\{a_k\}$ and $\{b_k\}$ are computed by carrying out the following integrations over the period T, say from $(-T/2)$ to $(T/2)$ or from zero to T, that is,

$$a_k = \frac{2}{T} \int_0^T x(t) \cos 2\pi f_k t \, dt \qquad k = 0, 1, 2, \ldots \qquad (1.11)$$

$$b_k = \frac{2}{T} \int_0^T x(t) \sin 2\pi f_k t \, dt \qquad k = 1, 2, 3, \ldots \qquad (1.12)$$

Note that

$$\frac{a_0}{2} = \frac{1}{T} \int_0^T x(t) \, dt = \mu_x \qquad (1.13)$$

where μ_x is the mean value of $x(t)$. Equations 1.10 through 1.13 are well known and can be put into other equivalent forms using $\omega = 2\pi f$ and $d\omega = 2\pi \, df$ in place of f. In this book, however, all results will be expressed using cyclical frequencies f in Hz rather than ω in radians per second.

Two alternate Fourier series formulas are also commonly used that follow directly from trigonometric and complex-valued relations. The first such formula is

$$x(t) = X_0 + \sum_{k=1}^{\infty} X_k \cos (2\pi f_k t - \theta_k) \qquad (1.14)$$

where

$$X_0 = \frac{a_0}{2}$$

$$X_k = \sqrt{a_k^2 + b_k^2} \qquad k = 1, 2, 3, \ldots$$

$$\theta_k = \tan^{-1}\left(\frac{b_k}{a_k}\right)$$

Here, $x(t)$ is expressed in a polar form rather than a rectangular form using amplitude factors $\{X_k\}$ and phase factors $\{\theta_k\}$ at the discrete frequencies f_k. The second formula is

$$x(t) = \sum_{k=-\infty}^{\infty} A_k e^{j2\pi f_k t} \qquad (1.15)$$

where

$$A_0 = \frac{a_0}{2}$$

$$A_k = \tfrac{1}{2}(a_k - jb_k) = \frac{1}{T} \int_0^T x(t) e^{-j2\pi f_k t} \, dt \qquad k = \pm 1, \pm 2, \pm 3, \ldots$$

This result is based on *Euler's relation* given by

$$e^{-j\theta} = \cos\theta - j\sin\theta \qquad (1.16)$$

Even though $x(t)$ may be real valued, it can be expressed in a theoretical complex-valued form using negative as well as positive frequency components. In particular, the factors $\{A_k\}$ will be complex valued with

$$A_k = |A_k|e^{-j\theta_k} \qquad k = \pm 1, \pm 2, \pm 3, \ldots \qquad (1.17)$$

where

$$|A_k| = \tfrac{1}{2}\sqrt{a_k^2 + b_k^2} = \frac{X_k}{2}$$

$$\theta_k = \tan^{-1}\left(\frac{b_k}{a_k}\right)$$

When $x(t)$ is real valued,

$$|A_{-k}| = |A_k| \qquad \theta_{-k} = -\theta_k$$
$$A_{-k} = |A_{-k}|e^{-j\theta_{-k}} = |A_k|e^{j\theta_k} = A_k^* \qquad (1.18)$$

where A_k^* is the complex conjugate of A_k. Examples of Fourier series are straightforward and are available from numerous books.

1.2.2   *Fourier Transforms*

Suppose the record $x(t)$ is nonperiodic as occurs for transient data (deterministic or random) or for stationary random data. The previous Fourier series representations can then be extended by considering what happens as T approaches infinity. This leads to the Fourier integral

$$X(f) = \int_{-\infty}^{\infty} x(t)e^{-j2\pi ft}\,dt \qquad -\infty < f < \infty \qquad (1.19)$$

where $X(f)$ will exist if

$$\int_{-\infty}^{\infty} |x(t)|\,dt < \infty \qquad (1.20)$$

The quantity $X(f)$ defined by Equation 1.19 is called the *direct Fourier transform* (or spectrum) of $x(t)$. Conversely, if $X(f)$ is known, then the *inverse Fourier transform* of $X(f)$ will give $x(t)$ by the formula

$$x(t) = \int_{-\infty}^{\infty} X(f)e^{j2\pi ft}\,df \qquad -\infty < t < \infty \qquad (1.21)$$

The associated $x(t)$ and $X(f)$ in Equations 1.19 and 1.21 are said to be *Fourier transform pairs*. Observe that $X(f)$ is generally a complex-valued function of both positive and negative frequencies, even when $x(t)$ is real valued. In terms of real and imaginary parts,

$$X(f) = X_R(f) - jX_I(f) \qquad (1.22)$$

where

$$X_R(f) = |X(f)| \cos \theta(f) = \int_{-\infty}^{\infty} x(t) \cos 2\pi ft \, dt$$

$$X_I(f) = |X(f)| \sin \theta(f) = \int_{-\infty}^{\infty} x(t) \sin 2\pi ft \, dt$$

In terms of complex polar notation, which will be used throughout this book,

$$X(f) = |X(f)|e^{-j\theta(f)} \qquad (1.23)$$

where $|X(f)|$ is the magnitude spectrum and $\theta(f)$ is the phase spectrum. Illustrations of Fourier spectra of three transient time history records are shown in Figure 1.5.

1.2.3   *Finite Fourier Transforms*

For a stationary random time history record $x(t)$, which theoretically exists over all time, the integral

$$\int_{-\infty}^{\infty} |x(t)| \, dt = \infty \qquad (1.24)$$

Hence its Fourier transform as given by Equation 1.19 will *not* exist. However, one cannot measure in the field or in the laboratory any $x(t)$ from $-\infty$ to $+\infty$. Instead, one can measure $x(t)$ only over some finite time interval T so that $X(f)$ is estimated by computing the *finite Fourier transform*

$$X_T(f) = X(f, T) = \int_0^T x(t)e^{-j2\pi ft} \, dt \qquad (1.25)$$

Such finite Fourier transforms will *always* exist for finite length records of stationary random data.

Equations 1.15 and 1.25 demonstrate that at the discrete frequencies $f_k = (k/T)$, the finite Fourier transform yields

$$X(f_k, T) = TA_k \qquad k = \pm 1, \pm 2, \pm 3, \ldots \qquad (1.26)$$

$$x(t) = \begin{cases} Ae^{-at} & t \geqslant 0 \\ 0 & t < 0 \end{cases} \qquad x(t) = \begin{cases} Ae^{-at} \cos bt & t \geqslant 0 \\ 0 & t < 0 \end{cases} \qquad x(t) = \begin{cases} A & 0 \leqslant t \leqslant c \\ 0 & c < t < 0 \end{cases}$$

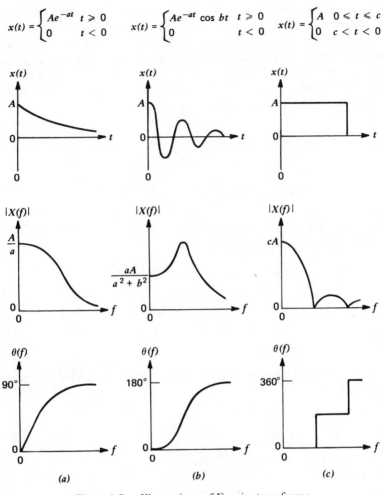

Figure 1.5   Illustrations of Fourier transforms.

Hence if f is restricted to take on only these discrete frequencies, then the finite Fourier transform calculations will actually produce a Fourier series of period T. For digital processing of data, this is precisely what occurs.

To be specific, one should be aware of digital procedures which are discussed at greater length in Reference 1.3. When $x(t)$ is sampled at points Δt apart, the record length becomes $T = N\Delta t$, where N is the sample size. This automatically induces a Nyquist cutoff frequency $f_c = (1/2\Delta t)$. Also, the computations treat the data as if they were periodic data of period T. Hence the fundamental frequency is $f_1 = (1/T)$ and results are obtained only at discrete frequencies spaced $\Delta f = f_1$ apart. The continuous record $x(t)$ is replaced by the data sequence $\{x_n\} = \{x(n\Delta t)\}$ for $n = 1, 2, 3, \ldots, N$, and the continuous Fourier transform $X(f)$ is replaced by the discrete Fourier transform sequence $\{X_k\} = \{X(k\Delta f)\}$ for $k = 1, 2, 3, \ldots, N$. Values beyond $k = (N/2)$ can be computed from earlier values since $f_c = (N/2)\Delta f$. The appropriate Fourier transform pair formulas are

$$X_k = X(k\Delta f) = \Delta t \sum_{n=1}^{N} x_n \exp\left(-j2\pi \frac{kn}{N}\right) \qquad k = 1, 2, 3, \ldots, N \quad (1.27)$$

$$x_n = x(n\Delta t) = \Delta f \sum_{k=1}^{N} X_k \exp\left(j2\pi \frac{kn}{N}\right) \qquad n = 1, 2, 3, \ldots, N \quad (1.28)$$

A few consequences of these formulas for real-valued sequences $\{x_n\}$ follow. Let X_k^* be the complex conjugate of X_k. Then

$$
\begin{aligned}
X_{-k} &= X_k^* && \text{for all } k \\
X_{N-k} &= X_k^* && \text{for } k = 1, 2, 3, \ldots, (N/2) \\
X_{(N/2)+k} &= X_{(N/2)-k}^* && \text{for } k = 1, 2, 3, \ldots, (N/2) \quad (1.29) \\
X_{k+N} &= X_k && \text{for all } k \\
x_{n+N} &= x_n && \text{for all } n
\end{aligned}
$$

Thus x_n is a repetitive function of n modulo N, and X_k is a repetitive function of k modulo N.

1.2.4 Delta Functions

Consider a rectangular-shaped function $f(t)$ with a magnitude $1/w$ and a width w centered at $t = 0$, as shown in Figure 1.6. The equation of this function is

$$
\begin{aligned}
f(t) &= \frac{1}{w} && -\frac{w}{2} \le t \le \frac{w}{2} \\
&= 0 && |t| > \frac{w}{2}
\end{aligned}
\quad (1.30)
$$

and the area under the function is given by the integral of $f(t)$ as follows:

$$A = \int_{-\infty}^{\infty} f(t)\, dt = \int_{-w/2}^{w/2} \frac{dt}{w} = \left(\frac{1}{w}\right)w = 1 \quad (1.31)$$

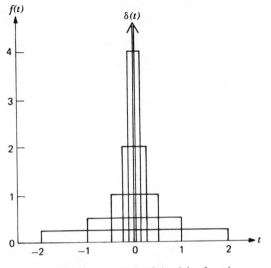

Figure 1.6 Evolution of the delta function.

Now let the base of $f(t)$ become increasing small with a corresponding increase in height so as to maintain unity area, as shown in Figure 1.6. In the limit as $w \to 0$, it follows that

$$
\delta(t) = \lim_{w \to 0} f(t) = \infty \qquad t = 0
$$
$$
= 0 \qquad t \neq 0 \tag{1.32}
$$

but the integral of the function, as given by Equation 1.31, remains unity, that is,

$$
\int_{-\varepsilon}^{\varepsilon} \delta(t) \, dt = \lim_{w \to 0} \left(\frac{w}{w} \right) = 1 \tag{1.33}
$$

where ε is an arbitrarily small value.

Limiting functions of this type are called *Delta functions* and are denoted by $\delta(t)$. Delta functions can be positioned at any point t_0 and can have any area A using the notation $A\delta(t - t_0)$. Specifically,

$$
A\delta(t - t_0) = \infty \qquad t = t_0
$$
$$
= 0 \qquad t \neq t_0 \tag{1.34}
$$

$$
\int_{t_0 - \varepsilon}^{t_0 + \varepsilon} A\delta(t - t_0) \, dt = A \tag{1.35}
$$

Furthermore, when any analytic function $x(t)$ is multiplied by a delta function $\delta(t - t_0)$ and integrated, the result is the value of $x(t)$ at $t = t_0$, that is,

$$\int_{-\infty}^{\infty} x(t)\delta(t - t_0)\, dt = x(t_0) \tag{1.36}$$

Thus the delta function can help pick out particular values of $x(t)$ at $t = t_0$.

Table 1.1

Special Fourier Transform Pairs

$x(t)$	$X(f)$
1	$\delta(f)$
$e^{j2\pi f_0 t}$	$\delta(f - f_0)$
$x(t - \tau_0)$	$X(f)e^{-j2\pi f \tau_0}$
$\cos 2\pi f_0 t$	$\frac{1}{2}[\delta(f - f_0) + \delta(f + f_0)]$
$\sin 2\pi f_0 t$	$\dfrac{1}{2j}[\delta(f - f_0) - \delta(f + f_0)]$
$\begin{aligned}&1;\ 0 \le t \le T\\&0;\ \text{otherwise}\end{aligned}$	$T\left(\dfrac{\sin \pi f t}{\pi f t}\right)e^{-j\pi f T}$
$2aB\left(\dfrac{\sin 2\pi B t}{2\pi B t}\right)$	$\begin{aligned}&a;\ -B \le f \le B\\&0;\ \text{otherwise}\end{aligned}$
$aB\left(\dfrac{\sin \pi B t}{\pi B t}\right)\cos 2\pi f_0 t$	$\begin{aligned}&a;\ f_0 - \dfrac{B}{2} \le f \le f_0 + \dfrac{B}{2}\\[4pt]&0;\ \text{otherwise}\end{aligned}$
$e^{-a\lvert t\rvert};\ a > 0$	$\dfrac{2a}{a^2 + (2\pi f)^2}$
$e^{-a\lvert t\rvert}\cos 2\pi f_0 t;\ a > 0$	$\dfrac{a}{a^2 + 4\pi^2(f + f_0)^2} + \dfrac{a}{a^2 + 4\pi^2(f - f_0)^2}$
$\displaystyle\int_{-\infty}^{\infty} x_1(u)\, x_2(t - u)\, du$	$X_1(f)\, X_2(f)$

The direct Fourier transform $X(f)$ associated with a delta function $\delta(t)$ is given by

$$X(f) = \int_{-\infty}^{\infty} \delta(t)e^{-j2\pi ft}\,dt = 1 \qquad \text{for all } f \qquad (1.37)$$

The inverse Fourier transform relation gives

$$x(t) = \int_{-\infty}^{\infty} e^{j2\pi ft}\,df = \delta(t) \qquad (1.38)$$

1.2.5   *Special Fourier Transforms*

A few examples of special Fourier transform pairs satisfying Equations 1.19 and 1.21 are listed in Table 1.1. Many other such pairs are readily available in published books on the subject.

1.3   PHYSICAL SYSTEM RESPONSE PROPERTIES

Most of the applications of correlation and spectral density functions considered in this book involve some physical system. A brief review is presented here of the response properties of physical systems that are pertinent to material in later chapters. The emphasis is on mechanical systems since they are the basis for most of the later illustrations. However, using classical analogies [1.3], the relationships presented here apply equally well to many other physical systems.

1.3.1   *Unit Impulse Response Functions*

An ideal physical system is one which (*a*) is physically realizable, (*b*) has constant parameters, (*c*) is stable, and (*d*) is linear, all to be defined shortly. For such an ideal system, the basic response properties of primary interest are given by the response of the system to a delta function input, called the *unit impulse response function* or the *weighting function* $h(\tau)$. Specifically, consider a system with a well-defined input $x(t)$ producing a well-defined output $y(t)$, as shown in Figure 1.7. The unit impulse response function is given by

$$h(t) = y(t) \qquad \text{when } x(t) = \delta(t) \qquad (1.39)$$

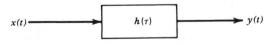

Figure 1.7   Ideal single input/single output system.

where t is time measured from the instant the delta function input is applied. The importance of the unit impulse response function as a description of the system is due to the following fact: For any arbitrary input $x(t)$, the system output $y(t)$ is given by the *superposition* or *convolution* integral

$$y(t) = \int_{-\infty}^{\infty} h(\tau)x(t - \tau)\, d\tau \tag{1.40}$$

That is, the response $y(t)$ is given by a weighted linear sum over the entire time history of the input $x(t)$.

A *physically realizable* system cannot respond to an input until that input has been applied. This requires that

$$h(\tau) = 0 \qquad \text{for } \tau < 0 \tag{1.41}$$

Hence for physically realizable systems, the lower limit of integration in Equation 1.40 is zero rather than minus infinity.

A physical system is said to have *constant parameters* if the unit impulse response function is not dependent on the time an input is applied, that is,

$$h(t, \tau) = h(\tau) \qquad \text{for } -\infty < t < \infty \tag{1.42}$$

If a physical system has constant parameters, stationary inputs will always produce stationary responses (after switch-on transients decay).

A physical system is said to be *stable* if every possible bounded input produces a bounded response. This condition is assured if

$$\int_{-\infty}^{\infty} |h(\tau)|\, d\tau < \infty \tag{1.43}$$

System stability is required for all the input/output relationships and applications discussed in this book.

A *linear* system is additive and homogeneous. Given two inputs x_1 and x_2 which individually produce outputs y_1 and y_2 per Equation 1.40, the system is *additive* if the input $x_1 + x_2$ produces an output $y_1 + y_2$, and is *homogeneous* if the input cx_1 produces the output cy_1, where c is an arbitrary constant. This essentially means that $h(\tau)$ is not dependent on the input, that is,

$$y(t) = \int_{0}^{\infty} h(\tau)x(t - \tau)\, d\tau \qquad \text{for all } x(t) \tag{1.44}$$

If a system is linear, random inputs with a Gaussian probability distribution (to be defined in Chapter 2) will produce outputs that also have a Gaussian probability distribution.

Linearity is the most likely property of physical systems to be violated in practice. In particular, with random inputs, there is usually a small probability of an instantaneous input so extreme that the system can no longer respond in a directly proportional manner as demanded by the homogeneous requirement. This is an important and difficult problem in those applications which involve extreme value statistics, for example, the prediction of catastrophic failures of structures under random loading. However, for most other applications, unless the system in question is strongly nonlinear, the correlation and coherence analysis procedures discussed in this book will yield meaningful results in terms of best linear approximations in the least-squares sense.

1.3.2 *Frequency Response Functions*

The dynamic properties of physical systems are usually described in terms of some linear transformation of the unit impulse response function $h(\tau)$ rather than $h(\tau)$ itself. Any one of several such linear transformations might be employed for special applications. For ideal systems, however, a Fourier transformation producing a direct frequency domain description of the system properties is most desirable from the viewpoint of the applications of concern in this book. The *Fourier* transform of the unit impulse response function where $h(\tau) = 0$ for $\tau < 0$ is given by

$$H(f) = \int_0^\infty h(\tau)e^{-j2\pi f\tau}\,d\tau \tag{1.45}$$

and is called the *frequency response function*. The frequency response function is generally a complex number with real and imaginary parts given by

$$H(f) = H_R(f) - jH_I(f) \tag{1.46}$$

$$H_R(f) = \int_0^\infty h(\tau)\cos 2\pi f\tau\,d\tau \qquad H_I(f) = \int_0^\infty h(\tau)\sin 2\pi f\tau\,d\tau$$

For convenience, it will be described throughout this book in terms of complex polar notation as follows:

$$H(f) = |H(f)|e^{-j\phi(f)} \tag{1.47}$$

$$|H(f)| = [H_R^2(f) + H_I^2(f)]^{1/2} \qquad \phi(f) = \tan^{-1}\left[\frac{H_I(f)}{H_R(f)}\right]$$

The magnitude $|H(f)|$ is commonly referred to as the system *gain factor* and the phase $\phi(f)$ is called the system *phase factor*. Note that the phase factor is defined so that lag angles are positive, to be consistent with the sign conventions used throughout this book.

The physical interpretation of the frequency response function is straightforward. For an ideal system as described in Section 1.3.1, a sinusoidal input at frequency f will produce a sinusoidal output at exactly the same frequency f. However, the amplitude of the output will generally be different from the input amplitude, and the output will generally be shifted in phase from the input as follows:

$$x(t) = X \sin 2\pi ft \qquad y(t) = Y \sin (2\pi ft - \theta) \qquad (1.48)$$

The ratio of the output to input amplitudes equals the system gain factor, and the phase shift between the output and input is the system phase factor at frequency f, that is,

$$H(f) = \frac{Y(f)}{X(f)} \qquad \phi(f) = \theta(f) \qquad (1.49)$$

The term *transfer function* is commonly used by engineers to denote the same quantity as the frequency response function of Equation 1.45. However, to be precise, the transfer function should be defined by the *Laplace* transform equation

$$H_1(p) = \int_0^\infty h(\tau)e^{-p\tau}\, d\tau \qquad p = a + jb \qquad (1.50)$$

where the real part of p, namely a, is not restricted to be zero. For $a \neq 0$, such Laplace transforms will be different from the Fourier transforms of Equation 1.45. Along the imaginary axis where $a = 0$, by taking $b = 2\pi f$, one obtains $H_1(j2\pi f) = H(f)$. Thus along the imaginary axis, the transfer function is the same as the frequency response function, which helps explain why these terms are often interchanged.

1.3.3   *Single Degree-of-Freedom System*

To illustrate a frequency response function of common interest, consider the single degree-of-freedom mechanical system consisting of a mass, spring, and dashpot, as shown in Figure 1.8. Assume the mass is subjected to a force input $F(t)$ producing a displacement response $y(t)$. From Newton's laws, the differential equation of motion describing the response of this system is given by

$$m\frac{d^2 y(t)}{dt^2} + c\frac{dy(t)}{dt} + ky(t) = F(t) \qquad (1.51)$$

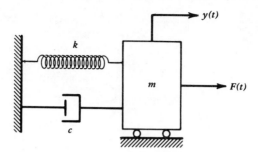

Figure 1.8   Single degree-of-freedom mechanical system.

To find the frequency response function of the system, let the input $F(t) = \delta(t)$, a delta function as defined in Section 1.2.4. Then from Equation 1.39, the response $y(t) = h(t)$, and from Equation 1.45, the Fourier transform of the response $Y(f) = H(f)$. Taking Fourier transforms of both sides of Equation 1.51 yields

$$[-(2\pi f)^2 m + j2\pi fc + k]Y(f) = 1$$

Thus

$$Y(f) = H(f) = [k - (2\pi f)^2 m + j2\pi fc]^{-1} \tag{1.52}$$

It is convenient to write Equation 1.52 in a different form by introducing two definitions:

$$\zeta = \frac{c}{2\sqrt{km}} \qquad f_n = \frac{1}{2\pi}\sqrt{\frac{k}{m}} \tag{1.53}$$

The term ζ in Equation 1.53 is called the *damping ratio* of the system and describes the system damping as a fractional portion of critical damping c_c. If the mass is displaced from its neutral position and released, c_c is that value of damping where the mass will just return to its neutral position without further oscillation; for the system in Figure 1.8, $c_c = 2\sqrt{km}$. The term f_n in Equation 1.53 is called the *undamped natural frequency* of the system. If the system had no damping and the mass were displaced from its neutral position and released, the system would perpetually oscillate at the frequency f_n. By using the definitions in Equation 1.53, the frequency response of the system in Equation 1.52 may be written as

$$H(f) = \frac{1/k}{1 - (f/f_n)^2 + j2\zeta f/f_n} \tag{1.54}$$

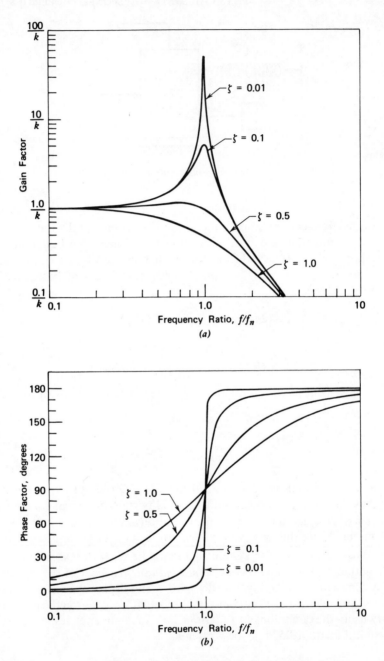

Figure 1.9 Frequency response function of a single degree-of-freedom system, force input and displacement output. (*a*) Gain factor. (*b*) Phase factor.

20

In terms of the system gain and phase factors defined in Equation 1.47,

$$|H(f)| = \frac{1/k}{\sqrt{[1 - (f/f_n)^2]^2 + [2\zeta f/f_n]^2}} \qquad (1.55a)$$

$$\phi(f) = \tan^{-1}\left[\frac{2\zeta f/f_n}{1 - (f/f_n)^2}\right] \qquad (1.55b)$$

Plots of these gain and phase factors are shown in Figure 1.9.

Two characteristics of the plots in Figure 1.9 are of particular interest. First, the gain factor has a peak at some frequency less than f_n for all cases where $\zeta \leq 1/\sqrt{2}$. The frequency at which this peak gain factor occurs is called the *resonance frequency* of the system. Specifically, it can be shown by minimizing the denominator of $|H(f)|$ in Equation 1.55a that the resonance frequency, denoted by f_r, is given by

$$f_r = f_n\sqrt{1 - 2\zeta^2} \qquad \zeta^2 \leq 0.5 \qquad (1.56)$$

and that the peak value of the gain factor which occurs at the resonance frequency is given by

$$|H(f_r)| = \frac{1/k}{2\zeta\sqrt{1 - \zeta^2}} \qquad \zeta^2 \leq 0.5 \qquad (1.57)$$

Second, the phase factor varies from $0°$ for frequencies much less than f_n to $180°$ for frequencies much greater than f_n. The exact manner in which $\phi(f)$ varies between these phase angle limits depends on the damping ratio ζ. However, for all values of ζ, the phase $\phi(f) = 90°$ for $f = f_n$.

Actual physical systems often have very small values of damping such that $\zeta \ll 1$. For example, mechanical structures generally have damping ratios of $\zeta < 0.05$. Hence it is common in practice to find physical systems with gain factors that display very sharp peaks and phase factors that show rapid $180°$ phase shifts. Such systems appear, in effect, to be narrow bandpass filters, and their bandwidth is commonly measured in terms of the *half-power point bandwidth* of the gain factor given by

$$B_r = f_2 - f_1 \qquad \text{where } |H(f_1)|^2 = |H(f_2)|^2 = \tfrac{1}{2}|H(f_r)|^2 \qquad (1.58)$$

For the usual case where the damping ratio is relatively small, it can be shown by substituting Equation 1.58 into Equation 1.55a that

$$B_r \approx 2\zeta f_r \qquad (1.59)$$

1.3.4 *Distributed Systems*

The parameters of physical systems are generally distributed in space. Hence the systems will have more than one frequency of resonance. For example, consider the uniform cantilever beam shown in Figure 1.10. If constant amplitude sinusoidal excitation is applied to this beam with increasing frequency starting from zero Hz, a frequency will be reached where the response at all points on the beam will achieve a maximum. As the frequency of excitation continues to increase, the beam response will at first diminish and then increase to a second maximum, and so on. Each of these response maxima corresponds to a resonance associated with a *normal mode* of the beam vibration. Assuming zero damping, the frequency of each resonance is called the *normal mode frequency*. If the excitation is removed after establishing a resonance and the undamped beam is permitted to vibrate

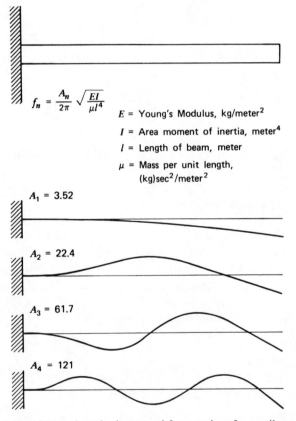

$$f_n = \frac{A_n}{2\pi} \sqrt{\frac{EI}{\mu l^4}}$$

E = Young's Modulus, kg/meter^2

I = Area moment of inertia, meter^4

l = Length of beam, meter

μ = Mass per unit length, (kg)sec^2/meter^2

A_1 = 3.52

A_2 = 22.4

A_3 = 61.7

A_4 = 121

Figure 1.10   Normal mode shapes and frequencies of a cantilever beam.

freely, the instantaneous deflection of the beam is called the *normal mode shape*. Theoretically, the beam will have an infinite number of such normal modes with associated mode shapes and frequencies. The first four normal mode shapes and frequencies for the uniform cantilever beam are shown in Figure 1.10.

For relatively simple structures like a cantilever beam, the normal mode shapes and frequencies can be determined mathematically by solving the appropriate partial differential equation describing the response [1.4]. For more complex structures, however, direct mathematical solutions are usually intractable and computer modeling procedures, such as finite element techniques, are commonly used to establish normal mode characteristics [1.5]. For an existing structure where excitations (either natural or artificial) can be applied, the normal mode characteristics may be established by appropriate data analysis procedures based on the principles detailed in Chapter 5 and Section 7.4 of this book.

Given the normal mode shapes and frequencies of a structure, the response of the structure at any spatial location ξ and time t is theoretically given by

$$y(\xi, t) = \sum_i \phi_i(\xi) q_i(t) \quad \cdot \quad i = 1, 2, 3, \ldots \tag{1.60}$$

where $\phi_i(\xi)$ is the mode shape of the ith normal mode and $q_i(t)$ is the generalized coordinate describing the response of the ith normal mode. The generalized coordinate of each mode is assumed to satisfy the relationship

$$M_i \frac{d^2 q_i(t)}{dt^2} + C_i \frac{dq_i(t)}{dt} + K_i q_i(t) = F_i(t) \tag{1.61}$$

where $M_i = \int_0^l \phi_i^2(\xi) m(\xi) \, d\xi$ = generalized mass

$C_i = \int_0^l \phi_i^2(\xi) c(\xi) \, d\xi$ = generalized damping

$K_i = 4\pi^2 f_i^2 M_i$ = generalized stiffness

$F_i(t) = \int_0^l \phi_i(\xi) p(\xi, t) \, d\xi$ = generalized force

In Equation 1.61, $m(\xi)$ is the mass density and $c(\xi)$ is the damping density of the structure at location ξ, and $p(\xi, t)$ is the applied load density at location ξ and time t. Note that the generalized mass, damping, stiffness, and force defined in Equation 1.61 are all functions of the mode shape and therefore take on different values for every different normal mode. Furthermore, note that the form of Equation 1.61 is identical to Equation 1.51, meaning that the frequency response at any point ζ for a given normal mode is the same as for the single degree-of-freedom system described in Figure 1.9. It follows that the frequency response function of the structure at any point ζ appears as the properly weighted sum of the frequency responses of a collection of single degree-of-freedom systems with different natural frequencies. See Reference 1.4 for further details on the response characteristics of such systems.

REFERENCES

1.1 Yaglom, A. M., *Stationary Random Functions*, Prentice-Hall, New Jersey, 1962.

1.2 Papoulis, A., *Signal Analysis*, McGraw-Hill, New York, 1977.

1.3 Bendat, J. S., and Piersol, A. G., *Random Data: Analysis and Measurement Procedures*, Wiley-Interscience, New York, 1971.

1.4 Stokey, W. F., "Vibration of Systems Having Distributed Mass and Elasticity," Chapter 7, in *Shock and Vibration Handbook*, 2nd Edition, (C. M. Harris and C. E. Crede, Eds.), McGraw-Hill, New York, 1976.

1.5 Huebner, K. H., *The Finite Element Method for Engineers*, Wiley, New York, 1975.

CHAPTER 2

PROBABILITY FUNCTIONS
AND AMPLITUDE MEASURES

As explained in Chapter 1, if a physical phenomenon of interest is random, then each time history measurement $x(t)$ of that phenomenon represents a unique set of circumstances which is not likely to be repeated in other independent measurements of that same phenomenon. Hence, to completely define all properties of the phenomenon, it is necessary to conceptually think in terms of all the time history measurements $\{x(t)\}$ that might have been made, as illustrated, previously in Figure 1.1. For the usual case of engineering interest where the phenomenon produces continuous time history data, an infinite number of such conceptual measurements is required to fully describe the phenomenon; that is, $N \to \infty$. It follows that the instantaneous amplitude of the phenomenon at a specific time t_1 in the future or from a different experiment cannot be determined from an exact equation, but instead must be defined in probabilistic terms.

2.1   PROBABILITY FUNCTIONS

There are a number of ways that one can define probability, but from an engineering viewpoint, the most convenient definition of probability is in terms of the relative frequency of occurrences. Specifically, assume an experiment is repeated many times under identical conditions. Let A be an outcome of the experiment which is of special interest. If the experiment is repeated N times and the outcome A occurs $N[A]$ times, then the probability of A is defined as

$$\text{Prob}[A] = \lim_{N \to \infty} \frac{N[A]}{N} \qquad (2.1)$$

In other words, the probability of an outcome A is the fractional portion of experimental results with the attribute A in the limit as the experiment is repeated an infinite number of times.

2.1.1   *Probability Distribution Functions*

Referring again to the ensemble of measurements in Figure 1.1, assume there is a special interest in a measured value at some time t_1 that is ξ units or less, that is, $A = x(t_1) \leq \xi$. It follows from the definition in Equation 2.1 that the probability of this occurrence is

$$\text{Prob}[x(t_1) \leq \xi] = \lim_{N \to \infty} \frac{N[x(t_1) \leq \xi]}{N} \qquad (2.2)$$

where $N[x(t_1) \leq \xi]$ is the number of measurements with an amplitude less than or equal to ξ at time t_1. The probability statement in Equation 2.2 can be generalized by letting the amplitude ξ take on arbitrary values, as illustrated in Figure 2.1. The resulting function of ξ is called the *probability distribution function* of the random process $\{x(t)\}$ at time t_1, and is given by

$$P_x(\xi, t_1) = P(x, t_1) = \text{Prob}[x(t_1) \leq \xi] \qquad (2.3)$$

where the notation $P(x, t_1)$ will be used henceforth for convenience. The probability distribution function then defines the probability that the instantaneous value of $x(t)$ from a future experiment at time t_1 will be less than

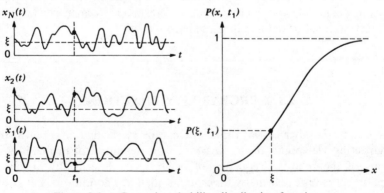

Figure 2.1   General probability distribution function.

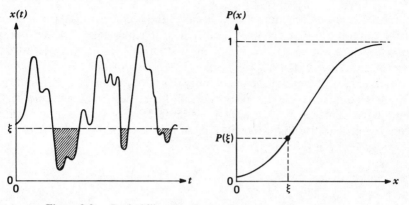

Figure 2.2   Probability distribution function of stationary data.

or equal to the amplitude ξ of interest. For the general case of nonstationary data, this probability will vary with the time t_1.

For the special case of stationary (ergodic) data, the probability distribution function will be the same at all times and can be determined from a single measurement $x(t)$ by

$$P(x) = \text{Prob}[x(t) \leq \xi] = \lim_{T \to \infty} \frac{T[x(t) \leq \xi]}{T} \qquad (2.4)$$

where $T[x(t) \leq \xi]$ is the total time that $x(t)$ is less than or equal to the amplitude ξ, as illustrated in Figure 2.2. In this case, the probability distribution function defines the probability that the instantaneous value of $x(t)$ from a future experiment at an arbitrary time will be less than or equal to any amplitude ξ of interest.

2.1.2   *Probability Density Functions*

Referring to Equations 2.2 and 2.3, since there must be some lower limit on the value of ξ (at minus infinity if not before) that $x(t)$ will always exceed, it follows that the value of the probability distribution function $P(x)$ will ultimately reach zero as ξ becomes small. Similarly, since there must be some upper value of ξ that $x(t)$ can never exceed, the value of $P(x)$ will ultimately reach unity as ξ becomes large. It is the manner in which this transition from zero to unity occurs that distinguishes data with different probability structures. Hence it is common to describe the probability structure of random data in terms of the slope of the distribution function given by the derivative

$$p(x, t_1) = \frac{dP(x, t_1)}{dx} \qquad (2.5)$$

where the resulting function $p(x, t_1)$ is called the *probability density function* of the random process $\{x(t)\}$ at time t_1. For the special case of stationary (ergodic) data, which is of primary interest in this book, there is no dependence on time and $p(x, t_1) = p(x)$.

Another way of viewing the probability density function is in terms of the way it is commonly measured. Specifically, consider a narrow amplitude window Δx centered at ξ, as shown in Figure 2.3. For stationary (ergodic) data, the probability that $x(t)$ will fall within this window at any arbitrary time is given from Equation 2.1 as

$$\text{Prob}[x(t) \in \Delta x_\xi] = \lim_{T \to \infty} \frac{T[x(t) \in \Delta x_\xi]}{T} \tag{2.6}$$

where the notation $\in$ means "within" and $T[x(t) \in \Delta x_\xi]$ is the time $x(t)$ falls within the window Δx centered at ξ. The probability density function is then obtained by dividing the probability in Equation 2.6 by Δx and taking the limit as $\Delta x \to 0$, that is,

$$p(x) = \lim_{\Delta x \to 0} \frac{\text{Prob}[x(t) \in \Delta x_\xi]}{\Delta x} \tag{2.7}$$

In other words, the probability density function is the rate of change of probability versus amplitude. It follows that probability statements are generated by integrating the probability density function, that is, calculating the area under $p(x)$ between any two amplitudes of interest, as illustrated

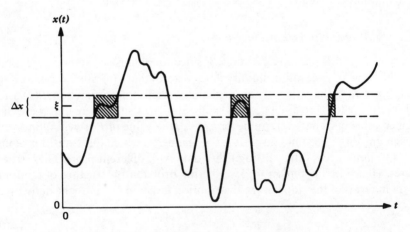

Figure 2.3 Probability density measurement.

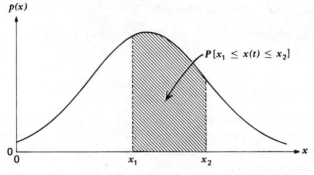

Figure 2.4   Probability calculation from density function.

in Figure 2.4. Specifically, the probability of finding an amplitude in the range between x_1 and x_2 is given by

$$\text{Prob}[x_1 < x(t) \le x_2] = \int_{x_1}^{x_2} p(x)\, dx = P(x_2) - P(x_1) \qquad (2.8)$$

By letting $x_1 \to -\infty$, it follows that

$$\text{Prob}[-\infty < x(t) \le x_2] = \int_{-\infty}^{x_2} p(x)\, dx = P(x_2) \qquad (2.9)$$

that is, the area under the probability density function below the amplitude x_2 equals the value of the probability distribution function evaluated at x_2. Furthermore, by letting $x_2 \to \infty$ as well, it follows that

$$\text{Prob}[-\infty < x(t) \le \infty] = \int_{-\infty}^{\infty} p(x)\, dx = 1 \qquad (2.10)$$

This shows the total area under the probability density function must be unity; that is, it is certain that the amplitude at any time will fall somewhere between plus and minus infinity.

2.2   MOMENTS AND AVERAGE AMPLITUDE MEASURES

For reasons to be demonstrated later, the instantaneous amplitude properties of random data can often be closely approximated through a knowledge of only two average properties that describe the central tendency and dispersion of the data. These average properties can be arrived at either by theoretical considerations or in terms of actual averaging operations on random time history records representing the data of interest.

2.2.1   *Expected Values and Moments*

Given any single valued function $g(x)$, the *expected value* of $g(x)$ is defined as

$$E[g(x)] = \int_{-\infty}^{\infty} g(x)p(x)\,dx \qquad (2.11)$$

In other words, the expected value of $g(x)$ is the average of $g(x)$ weighted by the likelihood of x occurring, as given by the probability density function of x. The expected value operation is linear, meaning it is additive and homogeneous as follows:

$$\begin{aligned} E[g(x) + h(x)] &= E[g(x)] + E[h(x)] \\ E[cg(x)] &= cE[g(x)] \end{aligned} \qquad (2.12)$$

The *moments* of a stationary random process $\{x(t)\}$ representing a physical phenomenon of interest are defined as

$$\mu_k = E[x^k] = \int_{-\infty}^{\infty} x^k p(x)\,dx \qquad k = 0, 1, 2, \ldots \qquad (2.13)$$

where $p(x)$ is the probability density function of $\{x(t)\}$ and μ_k is called the kth moment. For the zero moment ($k = 0$), it is clear that

$$\mu_0 = E[x^0] = \int_{-\infty}^{\infty} x^0 p(x)\,dx = 1 \qquad (2.14)$$

which is a trivial case. The first moment ($k = 1$) yields

$$\mu_1 = E[x^1] = \int_{-\infty}^{\infty} x p(x)\,dx = \mu \qquad (2.15)$$

which is called the *mean value* of $\{x(t)\}$ and henceforth will be denoted by μ. The second moment ($k = 2$) gives

$$\mu_2 = E[x^2] = \int_{-\infty}^{\infty} x^2 p(x)\,dx = \psi^2 \qquad (2.16)$$

which is called the *mean square value* of $\{x(t)\}$ and henceforth will be denoted by ψ^2. The positive square root of the mean square value is called the root mean square or *rms value*. For second and higher moments, it is often convenient to calculate moments about the mean, referred to as *central moments*. The second central moment is given by

$$\mu_2^c = E[(x - \mu)^2] = \int_{-\infty}^{\infty} (x - \mu)^2 p(x)\,dx = \sigma^2 \qquad (2.17)$$

which is called the *variance* of $\{x(t)\}$ and henceforth will be denoted by σ^2. The positive square root of the variance is called the *standard deviation*. The calculation of moments using Equation 2.13 can be extended to as high an order of k as desired, but for random data, the first two moments will usually suffice for reasons to be discussed later.

2.2.2   Central Tendency and Dispersion

The mean value μ (the first moment) defines the central tendency of the random process $\{x(t)\}$, while the variance σ^2 (second central moment) defines the dispersion of $\{x(t)\}$. The mean square value ψ^2 (the second moment) provides a measure of both central tendency and dispersion. Specifically, from Equation 2.17

$$\begin{aligned} \sigma^2 &= E[(x - \mu)^2] = E[x^2 - 2\mu x + \mu^2] \\ &= E[x^2] - 2\mu E[x] + \mu^2 = \psi^2 - \mu^2 \end{aligned} \tag{2.18}$$

It follows that the mean square value is given by

$$\psi^2 = \sigma^2 + \mu^2 \tag{2.19}$$

Hence the variance and mean square value are equal for the special case where the mean value μ is zero.

It might be helpful to observe that the moments of random processes defined in Equations 2.15 through 2.17 are directly analogous to the moments commonly used to describe the distribution of mass in a structure. Specifically, given a structure with a mass density function $m(x)$, the central tendency of the mass, called the center of gravity (C.G.), is given by

$$\text{C.G.} = \frac{\int_{-\infty}^{\infty} xm(x)\,dx}{\int_{-\infty}^{\infty} m(x)\,dx} \tag{2.20}$$

while the dispersion of the mass, called the radius of gyration squared, is defined as

$$K^2 = \frac{\int_{-\infty}^{\infty} x^2 m(x)\,dx}{\int_{-\infty}^{\infty} m(x)\,dx} \tag{2.21}$$

The radius of gyration squared about the C.G. is given by

$$K_0^2 = \frac{\int_{-\infty}^{\infty} (x - \text{C.G.})^2 m(x)\,dx}{\int_{-\infty}^{\infty} m(x)\,dx} \tag{2.22}$$

Finally, by algebra similar to that in Equation 2.18, it can be shown that

$$K^2 = K_0^2 + (\text{C.G.})^2 \tag{2.23}$$

By substituting probability density $p(x)$ for mass density $m(x)$, Equations 2.20 through 2.23 become identical to Equations 2.15 through 2.18, respectively. Hence the mean value, rms value, and standard deviation of a random process are directly analogous to the C.G., radius of gyration, and radius of gyration about the C.G. of a structural mass.

2.2.3   Data Averaging Operations

The mean value, mean square value, and variance defined in Equations 2.15 through 2.17, respectively, can be arrived at by more direct operations on the time history data representing the random process $\{x(t)\}$. Specifically, referring back to the ensemble of records shown in Figure 1.1, for the general case of nonstationary data, the mean value at a specific time t_1 is given by

$$\mu(t_1) = E[x(t_1)] = \lim_{N \to \infty} \frac{1}{N} \sum_{i=1}^{N} x_i(t_1) \qquad (2.24)$$

In other words, one simply calculates the average amplitude at the specific time t_1 over the ensemble of N measurements in Figure 1.1 ideally in the limit as $N \to \infty$. Similarly, the mean square value is obtained by calculating the average squared amplitude at time t_1 over the ensemble, that is,

$$\psi^2(t_1) = E[x^2(t_1)] = \lim_{N \to \infty} \frac{1}{N} \sum_{i=1}^{N} x_i^2(t_1) \qquad (2.25)$$

and the variance is computed by the same operations except the mean value is subtracted from the data before taking the square, specifically,

$$\sigma^2(t_1) = E[\{x(t_1) - \mu(t_1)\}^2] = \lim_{N \to \infty} \frac{1}{N} \sum_{i=1}^{N} \{x_i(t_1) - \mu(t_1)\}^2 \quad (2.26)$$

For the special case of stationary (ergodic) data, the various moments are constants independent of time and furthermore can be calculated from a single time history measurement $x(t)$ as follows:

$$\mu = E[x(t)] = \lim_{T \to \infty} \frac{1}{T} \int_0^T x(t) \, dt \qquad (2.27)$$

$$\psi^2 = E[x^2(t)] = \lim_{T \to \infty} \frac{1}{T} \int_0^T x^2(t) \, dt \qquad (2.28)$$

$$\sigma^2 = E[\{x(t) - \mu\}^2] = \lim_{T \to \infty} \frac{1}{T} \int_0^T \{x(t) - \mu\}^2 \, dt \qquad (2.29)$$

2.3   SPECIAL PROBABILITY DENSITY FUNCTIONS

The actual probability density functions associated with physical pheno-mena in practice are endless in number. For the purposes of the material in this book, however, an understanding of three special probability density functions which closely approximate a wide class of data of interest will suffice. These three are the probability density functions for (*a*) normal (Gaussian) noise, (*b*) sine waves, and (*c*) sine waves in noise. Since these three cases are well known, they are presented and discussed here without deriva-tions. See Reference 2.1 for details.

2.3.1   *Normal (Gaussian) Noise*

The data representing a wide collection of random physical phenomena in practice tend to have probability density functions that are closely ap-proximated by

$$p(x) = \frac{1}{\sigma\sqrt{2\pi}} e^{-(x-\mu)^2/2\sigma^2} \tag{2.30}$$

where μ and σ are the mean value and standard deviation, respectively, of the data. The function in Equation 2.30, which was first deduced in 1733 as the limit of the binomial probability function, is commonly called the *normal* or *Gaussian* probability density function. For convenience, it is usually tabulated and plotted in terms of the standardized variable

$$z = \frac{(x-\mu)}{\sigma} \tag{2.31}$$

which has zero mean and unity standard deviation. With this simple trans-formation of variables, Equation 2.30 becomes

$$p(z) = \frac{1}{\sqrt{2\pi}} e^{-z^2/2} \tag{2.32}$$

which is called the standardized normal or Gaussian probability density function. From Equation 2.9, the standardized normal probability distribu-tion function is given by

$$P(z) = \frac{1}{\sqrt{2\pi}} \int_{-\infty}^{z} p(\xi) \, d\xi \tag{2.33}$$

Both the probability density and distribution of standardized normal data are illustrated in Figure 2.5, and their values are tabulated in many text-books including Reference 2.2.

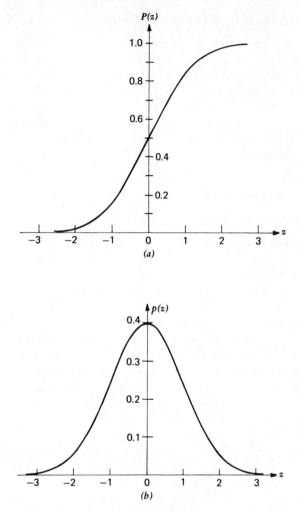

Figure 2.5 Standardized Gaussian probability distribution and density functions. (*a*) Distribution function. (*b*) Density function.

The importance of the normal distribution evolves from the practical implications of the Central Limit theorem in statistics, which may be broadly stated as follows: If a random variable x is in fact the net result of a linear sum of n statistically independent constituent variables, $x_1, x_2, \ldots, x_n$, then no matter what probability density functions the constituent variables may have, the probability density of $x = x_1 + x_2 + \cdots + x_n$ will approach the normal form given in Equation 2.30 as n approaches infinity. Since most random physical phenomena are really the net result of a number of constituent random events, the normal equation often provides a reasonable approximation to the probability density functions of random data.

It is very desirable to assume random data are normally distributed for two reasons. First, from Equation 2.30, the normal function is totally determined by the mean value μ and standard deviation σ. Hence only these two parameters need to be measured to determine the probability density function of the data. Second, there is a basic theorem [2.3] which states that all linear operations on a normally distributed random variable produce a new random variable which is also normally distributed. This means that various linear operations in mathematics, such as integration and the taking of Fourier transforms, can be performed freely on normally distributed data to obtain results which also will be normally distributed. In fact, if the linear operations are such that they reduce the frequency range of the data, they will tend to suppress any deviations from the Gaussian form that may exist, assuming no deterministic components are present.

As a word of caution, it should be noted that theoretical normality defined in Equation 2.30 is not bounded in amplitude, suggesting that there is a finite probability of exceeding any upper amplitude value no matter how large, or of not exceeding any lower amplitude value no matter how small. Since all physical phenomena and the measured data representing them must ultimately be limited in value by nonlinear restraints in both the positive and negative directions, it follows that no random data can be truly Gaussian. This fact becomes very important in those applications which involve an assessment of extreme values, for example, the prediction of an historically extreme and potentially catastrophic wind load or ocean wave height. A Gaussian assumption would be inappropriate in this case because wind load and wave height statistics deviate dramatically from the Gaussian form at extreme values. For most of the applications of concern in this book, however, a Gaussian assumption will suffice for the random data of interest, assuming it contains no deterministic components.

2.3.2   *Sine Waves*

The most common deterministic data are periodic data that can be decomposed into a collection of harmonically related sine waves. For a single sine wave by itself, a probabilistic distribution is not theoretically necessary since the exact amplitude at any future instant of time is given by $x(t) = X \sin(2\pi ft + \theta)$. However, if the phase angle θ is assumed to be a random variable with a uniform distribution between $\pm \pi$, then a sinusoidal function can be viewed as a random process, and the probability density function of the sinusoidal process, assuming a mean value of zero, can be shown to be [2.1]

$$p(x) = \left(\pi\sqrt{2\sigma^2 - x^2}\right)^{-1} \qquad |x| < X$$
$$= 0 \qquad\qquad\qquad |x| \geq X \qquad (2.34)$$

where $\sigma = X/\sqrt{2}$ is the standard deviation of the sine wave. A plot of this probability density function standardized for $\sigma = 1$ is shown in Figure 2.6. Also shown in Figure 2.6 is an illustration that helps explain why the sine wave probability density function has the "dish-shaped" characteristic. From Equation 2.7, the probablity density can be viewed in terms of a limiting operation on the probability that $x(t)$ falls within a window Δx, which in turn is the fractional portion of time that $x(t)$ falls within Δx. From Figure 2.6, it is clear that for any given window width Δx, a sine wave during each period spends the largest portion of time at the peak values $\pm X$, and the minimum portion of time at the mean value $\mu = 0$.

Like Gaussian random noise discussed in Section 2.3.1, the probability density function of a sine wave is determined completely from the mean and standard deviation of the data. Unlike Gaussian noise, however, the probability density reaches a minimum at the mean value of a sine wave, meaning a value at the mean is the least likely event between $\pm X$. From Figure 2.5, a value at the mean is the most likely event for Gaussian noise. This constitutes an important distinction between a sine wave and narrow-band noise which will generally be Gaussian no matter how narrow the noise bandwidth may become.

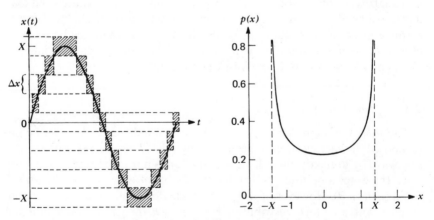

Figure 2.6   Standardized probability density function of a sine wave.

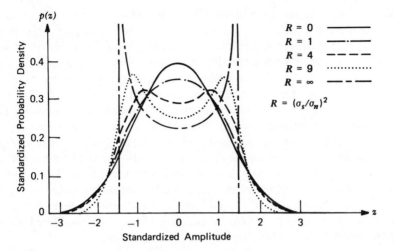

Figure 2.7   Standardized probability density function of a sine wave in Gaussian noise.

2.3.3   Sine Wave in Gaussian Noise

Now consider the case where a stationary (ergodic) random time history is of the form $x(t) = n(t) + s(t)$, where $n(t)$ is Gaussian random noise and $s(t) = S \sin (2\pi ft + \theta)$, a sine wave. The probability density function of $x(t)$ is simply the convolution of the individual density functions in Equations 2.30 and 2.34. Assuming zero mean values for both $n(t)$ and $s(t)$, the probability density can be shown to be [2.1].

$$p(x) = \frac{1}{\sigma_n \pi \sqrt{2\pi}} \int_0^\pi \exp\left[-\left(\frac{x - S \cos \theta}{4\sigma_n} \right)^2 \right] d\theta \qquad (2.35)$$

where σ_n is the standard deviation of the Gaussian noise $n(t)$, and S and θ are the amplitude and phase, respectively, of the sine wave. Plots of $p(x)$ for various ratios of sine wave variance to noise variance $R = \sigma_s^2/\sigma_n^2$ are shown in Figure 2.7.

2.4   STATISTICAL SAMPLING ERRORS

Referring back to Section 2.1.2, Equations 2.6 and 2.7 define the probability density function in a manner consistent with the way it is measured. Similarly, Equations 2.24 through 2.29 in Section 2.2.3 define the mean value, mean square value, and variance in measurement-oriented terms. All of these definitions involve limiting operations which cannot be carried out in practice; that is, it is clearly impossible to analyze an infinite number of records or a single record of infinite length. The inability to carry out such limiting operations leads to statistical sampling errors in the results of actual analyses. In other words, the result of an actual analysis of random data will constitute only a sample estimate of the true value of the parameter of interest. The magnitude of the potential error of this sample estimate is obviously important to the interpretations and applications of the analysis, hence it must be evaluated. It should be emphasized that the errors discussed here are those which arise from statistical sampling considerations in the data analysis procedures only. There are, of course, numerous other potential sources of error in the data acquisition procedures as noted earlier in Section 1.1.3. It is assumed that these data acquisition errors have been accounted for by appropriate calibration procedures.

2.4.1   *Bias versus Random Errors*

The errors that occur in the analysis of random data may be divided into two classes. The first is a haphazard scatter in the results from one analysis to the next of different samples of the same random data, and is called the *random error*. Random errors are a direct result of the fact that averaging operations must be performed over a finite number of sample records N or over a single sample record of finite length T. It follows that all analyses will involve a random error. The second type of error is a systematic error that will appear with the same magnitude and in the same direction from one analysis to the next, and is called the *bias error*. Bias errors generally evolve from windowing operations associated with the calculation of derivatives, for example, the Δx amplitude window in Equation 2.7 needed to convert a probability estimate to a probability density estimate. Theoretically, $p(x)$ in Equation 2.7 is defined in the limit as $\Delta x \to 0$, but in practice Δx must be finite.

To further clarify these two types of errors, consider a situation where a parameter ϕ is estimated repeatedly from independent sample records to obtain a collection of estimates $\hat{\phi}_i$, $i = 1, 2, 3, \ldots,$ as shown in Figure 2.8. The bias error of the estimate $\hat{\phi}$ is given by the expected value of the estimate (the average of repeated estimates) minus the value ϕ being estimated, that is,

$$\text{bias error} = b_{\hat{\phi}} = E[\hat{\phi}] - \phi = \lim_{N \to \infty} \frac{1}{N} \sum_{i=1}^{N} \hat{\phi}_i - \phi \tag{2.36}$$

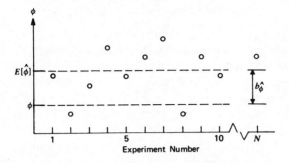

Figure 2.8 Random and bias errors in parameter estimates.

The random error of the estimate $\hat{\phi}$ is defined by the standard deviation of the estimate about its expected value, that is,

$$\text{random error} = \sigma_{\hat{\phi}} = \left[\lim_{N \to \infty} \frac{1}{N} \sum_{i=1}^{N} \{\hat{\phi}_i - E[\hat{\phi}]\}^2 \right]^{1/2} \qquad (2.37)$$

It is convenient to define errors in terms of a fractional portion of the parameter value being estimated. These normalized errors will be denoted by ε as follows:

$$\varepsilon_b = \frac{b_{\hat{\phi}}}{\phi} \qquad \varepsilon_r = \frac{\sigma_{\hat{\phi}}}{\phi} \qquad (2.38)$$

For example, if $\varepsilon_b = 0.1$, this would mean that the estimate $\hat{\phi}$ is, on the average, 10% greater than ϕ. If $\varepsilon_r = 0.1$, this would mean that the scatter in $\hat{\phi}$ about the average estimate would have a standard deviation of 10% of the value of ϕ. The normalized random error ε_r is often called the *coefficient of variation* of the estimate.

The probability density function of a sample estimate, commonly referred to as the *sampling distribution* of the estimate, can be very complicated in some cases. However, as a rule of thumb, if the random error is not too large, say $\varepsilon_r < 0.2$, then the sampling distribution can be approximated with reasonable accuracy by the normal distribution defined in Equation 2.30, where the mean value is $\mu_{\hat{\phi}} = (1 + \varepsilon_b)\phi$ and the standard deviation is $\sigma_{\hat{\phi}} = \varepsilon_r \phi$ as follows:

$$p(\hat{\phi}) \approx \frac{1}{\varepsilon_r \phi \sqrt{2\pi}} \exp\left\{ \frac{-[\hat{\phi} - (1 + \varepsilon_b)\phi]^2}{2(\varepsilon_r \phi)^2} \right\} \qquad (2.39)$$

2.4.2 *Errors in Amplitude Estimates*

There are no bias errors inherent in the estimate of mean and mean square values, assuming the estimates are correctly calculated using Equations 2.24 and 2.25 with a finite number of records N, or Equations 2.27 and 2.28 with a finite record length T in sec. The random errors in these estimates are well known and are proved in Reference 2.2 to be as summarized in Table 2.1. For ensemble averaged estimates, assuming statistical independence among all sample records, the error formulas in Table 2.1 are exact. For the time averaged estimates, however, the error formulas apply only to the situation where the energy of the data is uniformly distributed in frequency over a bandwidth of B in Hz. For this case, it can be shown from sampling theory [2.1] that $N \approx 2BT$. Since this situation rarely occurs, the error formulas for time averaged estimates constitute only a representation of the general form of the error and should not be used as a quantitative measure of the error.

There are two aspects of the error expressions in Table 2.1 that are of considerable importance because they are typical of the random error formulas for all parameter estimates to follow in this book. First, the random error ε_r is inversely proportional to the *square root* of the number of records N or record length T. Hence to cut ε_r in half, N or T must be increased fourfold. Second, for the time averaged estimates, the data bandwidth B is as important to the effective sample size of the estimate as the record length T. This means that for those applications where the frequency bandwidth of the data is very wide, as often occurs in communications problems, a rela-

Table 2.1

Random Errors in Mean and Mean Square Value Estimates

Estimation Procedure	Estimates and Corresponding Random Error, ε_r			
	Mean Value Estimates		Mean Square Value Estimates	
	Estimate, $\hat{\mu}_x$	Error, ε_r	Estimate, $\hat{\psi}_x^2$	Error, ε_r
Ensemble Averaging	$\dfrac{1}{N}\sum\limits_{i=1}^{N} x_i$	$\dfrac{\sigma_x}{\mu_x\sqrt{N}}$	$\dfrac{1}{N}\sum\limits_{i=1}^{N} x_i^2$	$\sqrt{\dfrac{2}{N}}$
Time Averaging	$\dfrac{1}{T}\displaystyle\int_0^T x(t)\,dt$	$\dfrac{\sigma_x}{\mu_x\sqrt{2BT}}$	$\dfrac{1}{T}\displaystyle\int_0^T x^2(t)\,dt$	$\dfrac{1}{\sqrt{BT}}$

tively short record might provide highly accurate estimates. On the other hand, for those applications where the bandwidth of the data is typically narrow, as occurs, for example, in atmospheric turbulence and ocean wave data, very long records may be required to obtain acceptably accurate results.

2.4.3   *Errors in Probability Density Estimates*

As mentioned earlier, the estimation of probability density functions using Equations 2.6 and 2.7 with a finite window Δx and record length T will involve both a bias and random error. The bias error evolves from the fact that Equation 2.7 with finite Δx will yield the average probability density over the amplitude range $x_0 \pm \Delta x/2$. For convenience of plotting, if nothing else, the estimate $\hat{p}(x_0)$ will generally be associated with the center amplitude of the window Δx. However, $\hat{p}(x_0)$ will generally not equal the true probability density $p(x_0)$ unless the first derivative of $p(x)$ with respect to x is a constant; that is, $p'(x) = dp(x)/dx = c$, a constant. This point is illustrated for an arbitrary probability density function in Figure 2.9. As a first order of approximation, the bias error in the estimate $\hat{p}(x)$ is given by [2.2]

$$\varepsilon_b[\hat{p}(x)] \approx \frac{(\Delta x)^2}{24} \frac{p''(x)}{p(x)} \tag{2.40}$$

where $p''(x)$ is the second derivative of $p(x)$ with respect to x.

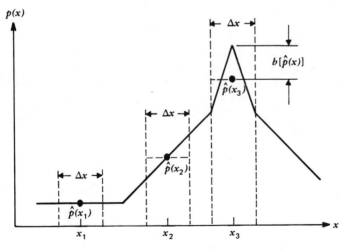

Figure 2.9   Bias error in probability density estimate.

For the case where $\hat{p}(x)$ is estimated by ensemble averaging over N statistically independent time history records, the random error in the estimate is approximated by [2.2]

$$\varepsilon_r[\hat{p}(x)] \approx \frac{1}{\sqrt{N \Delta x p(x)}} \qquad (2.41)$$

When $\hat{p}(x)$ is computed by time averaging a single sample record of length T with energy distributed uniformly over a bandwidth B, the random error is approximated by

$$\varepsilon_r[\hat{p}(x)] \approx \frac{1}{\sqrt{2BT \Delta x p(x)}} \qquad (2.42)$$

Like the error expressions for the time averaged estimates in Table 2.1, Equation 2.42 provides only an indication of the form of the error since little data have constant energy over a well-defined bandwidth B.

The results in Equations 2.41 and 2.42 reveal the same dependence of the random error on N, T, and B that appears in the error expressions for mean and mean square value estimates in Table 2.1. In this case, however, an additional term is present in the denominator of ε_r, namely, the amplitude window width Δx. Referring to Equation 2.40, Δx appears in the numerator of the bias error expression. Hence the selection of a value Δx in probability density analysis will always involve a compromise between random and bias errors. Of course, ε_r and ε_b can both be made as small as desired by simply increasing the record length T.

REFERENCES

2.1 Bendat, J. S., *Principles and Applications of Random Noise Theory*, Wiley, New York, 1958. Reprinted by Krieger Publishing Co., New York, 1977.

2.2 Bendat, J. S., and Piersol, A. G., *Random Data: Analysis and Measurement Procedures*, Wiley-Interscience, New York, 1971.

2.3 Papoulis, A., *Probability, Random Variables, and Stochastic Processes*, McGraw-Hill, New York, 1965.

CHAPTER 3

CORRELATION AND
SPECTRAL DENSITY FUNCTIONS

A wide range of engineering applications of random data analysis centers around the determination of linear relationships between two or more sets of data. These linear relationships are generally extracted in terms of a correlation function, or its Fourier transform called a spectral density function. Correlation and spectral density functions provide basically the same information, but from an historic viewpoint, they evolved separately; correlation functions were a product of mathematicians and statisticians, whereas spectral density functions were developed more directly as an engineering tool. For certain applications, correlation functions may provide the desired information in a more convenient format, but in recent years, spectral density functions have become the more common analysis tool for engineering problems.

3.1 CORRELATION FUNCTIONS

For purposes of introduction, it is helpful to view correlation functions in a classical context familiar to everyone who has had a course in elementary statistics. These classical ideas then readily extend to time history data by the introduction of a time delay parameter.

3.1.1 *Classical Correlation Concepts*

Consider an experiment producing two sets of paired measurements: $x_i, i = 1, 2, 3, \ldots, N$, and $y_i, i = 1, 2, 3, \ldots, N$. For example, x_1 is a load

43

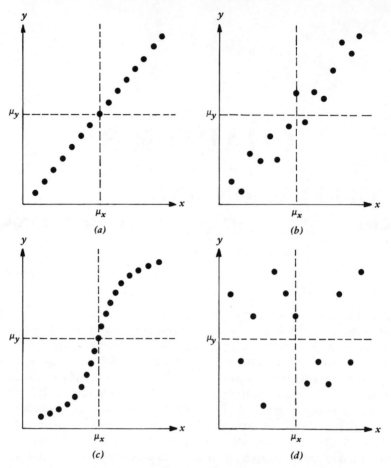

Figure 3.1 Varying degrees of correlations. (*a*) Perfect linear correlation. (*b*) Moderate linear correlation. (*c*) Nonlinear correlation. (*d*) No correlation.

applied to a structure and y_1 is the measured strain of the structure; x_2 and y_2 are a different load and resulting strain, respectively. Such load and strain measurements might be made for N different load levels to obtain the results shown in Figure 3.1. Ideally, the resulting measurements would be as shown in Figure 3.1*a* where there is a perfect linear relationship between the load x and the strain y. In this case, one can immediately write the equation

$$y = a + bx \tag{3.1}$$

where a is the zero intercept and b is the slope of the linear dependence of y on x. At the other extreme, if the strain y were improperly measured to produce only noise which was unrelated to the load, the results would be as shown in Figure 3.1d. In practice, something between these two extremes would probably be measured, as illustrated in Figures 3.1b and c. In Figure 3.1b, there is an underlying linear relationship between x and y, but not an analytically perfect one because of some randomness in the relationship and/or noise in the measurements. In Figure 3.1c, there is a perfect analytical relationship between x and y, but it is nonlinear.

The procedure for assessing the degree of linear dependence between x and y is straightforward; one simply calculates the average product of $x - \mu_x$ and $y - \mu_y$. In the limit as the sample size $N \to \infty$, this average product yields the *covariance* of x and y, defined as follows:

$$\sigma_{xy} = E[(x - \mu_x)(y - \mu_y)] = \lim_{N \to \infty} \frac{1}{N} \sum_{i=1}^{N} (x_i - \mu_x)(y_i - \mu_y) \quad (3.2)$$

For the case in Figure 3.1d, where x and y are unrelated, the sum of the positive products of $x_i - \mu_x$ and $y_i - \mu_y$ will equal the sum of the negative products and the average product will be $\sigma_{xy} = 0$. For the situation in Figure 3.1a, however, $y_i - \mu_y$ is always positive when $x_i - \mu_x$ is positive, and always negative when $x_i - \mu_x$ is negative, so the products are all positive yielding an average product of

$$\sigma_{xy} = \sigma_x \sigma_y \quad (3.3)$$

As will be demonstrated shortly, Equation 3.3 defines the largest possible value of the covariance between two random variables. Hence it is common to express correlation in terms of the ratio

$$\rho_{xy} = \frac{\sigma_{xy}}{\sigma_x \sigma_y} \qquad -1 \le \rho_{xy} \le 1 \quad (3.4)$$

where ρ_{xy} is called the *correlation coefficient*. The quantity ρ_{xy} can be used to assess the degree of linear dependence between any two variables x and y on a scale from $-$ unity to $+$ unity; $\rho_{xy} = -1$ simply means that the relationship between x and y is inverse rather than direct. The data in Figures 3.1b and c would be associated with a correlation coefficient between zero and $\pm$ unity.

The bounds on ρ_{xy} in Equation 3.4 may be proved as follows: For any real constants a and b, the expected value

$$E[\{a(x - \mu_x) + b(y - \mu_y)\}^2] \ge 0 \quad (3.5)$$

Expanding Equation 3.5 gives

$$E[a^2(x - \mu_x)^2 + 2ab(x - \mu_x)(y - \mu_y) + b^2(y - \mu_y)^2]$$
$$= a^2\sigma_x^2 + 2ab\sigma_{xy} + b^2\sigma_y^2 \ge 0 \quad (3.6)$$

Assuming $b \neq 0$, one can now divide through Equation 3.6 by b^2 to obtain

$$\left(\frac{a}{b}\right)^2 \sigma_x^2 + 2\left(\frac{a}{b}\right)\sigma_{xy} + \sigma_y^2 \geq 0 \tag{3.7}$$

which is a quadratic equation in (a/b) without real different roots since one side is non-negative. Therefore, the discriminant of this quadratic equation in a/b must be nonpositive, that is,

$$\text{discriminant} = 4\sigma_{xy}^2 - 4\sigma_x^2\sigma_y^2 \leq 0 \tag{3.8}$$

It follows that $\sigma_{xy}^2 \leq \sigma_x^2\sigma_y^2$, hence

$$-1 \leq \frac{\sigma_{xy}}{\sigma_x\sigma_y} \leq 1$$

as stated in Equation 3.4.

An important relationship involving the correlation coefficient arises in those problems where it can be assumed that y is a result of linear operations on x plus statistically independent noise n (which includes nonlinear effects) as follows:

$$y = a + bx + n \tag{3.9}$$

For such cases, the variance of y which is due to linear contributions of x alone, denoted by $\sigma_{y:x}^2$, is given by

$$\sigma_{y:x}^2 = \rho_{xy}^2\sigma_y^2 \tag{3.10}$$

In other words, the squared value of the correlation coefficient between x and y defines the fractional portion of the variance of y which can be attributed to the linear effects of x. It follows that the variance of y due to all other effects including nonlinearities will be

$$\sigma_{y:n}^2 = (1 - \rho_{xy}^2)\sigma_y^2 \tag{3.11}$$

The result in Equation 3.10, sometimes referred to as the *correlated output (power) relationship*, is proved as follows: From Equation 3.9, assuming the noise n has a zero mean value, the variance of y is given by

$$\begin{aligned}
\sigma_y^2 &= E[(y - \mu_y)^2] = E[\{b(x - \mu_x) + n\}^2] \\
&= E[b^2(x - \mu_x)^2 + 2b(x - \mu_x)n + n^2] = b^2\sigma_x^2 + \sigma_n^2
\end{aligned} \tag{3.12}$$

It is clear in Equation 3.12 that $b^2\sigma_x^2$ is that portion of the variance of y due to x and σ_n^2 is the portion due to n, that is,

$$\sigma_{y:x}^2 = b^2\sigma_x^2 \qquad \sigma_{y:n}^2 = \sigma_n^2 \tag{3.13}$$

Now, the covariance between x and y is given by Equation 3.2 as

$$\sigma_{xy} = E[(x - \mu_x)(y - \mu_y)] = E[(x - \mu_x)\{b(x - \mu_x)\}]$$
$$= bE[(x - \mu_x)^2] = b\sigma_x^2 \tag{3.14}$$

It follows from Equations 3.4 and 3.13 that

$$\rho_{xy}^2 \sigma_y^2 = \frac{b^2 \sigma_x^4}{\sigma_x^2 \sigma_y^2} \sigma_y^2 = b^2 \sigma_x^2 = \sigma_{y:x}^2$$

as stated in Equation 3.10. Furthermore, using Equation 3.12

$$(1 - \rho_{xy}^2)\sigma_y^2 = \sigma_y^2 - b^2 \sigma_x^2 = \sigma_{y:n}^2$$

as stated in Equation 3.11.

3.1.2   *Extentions to Time History Data*

Now consider the situation where the data of interest are measurements of two continuous random processes $\{x(t)\}$ and $\{y(t)\}$ which are assumed to be stationary (ergodic). Hence they can be represented by individual time history records $x(t)$ and $y(t)$. The classical concepts of correlation discussed in the preceding section apply directly, except now an additional variable is introduced, namely a time delay τ between $x(t)$ and $y(t)$. Referring back to Chapter 1, time may be replaced by any other independent variable of interest, for example, distance, in which case τ would be a spatial separation rather than a time delay.

From the definition in Equation 3.2, the *covariance function* between $x(t)$ and $y(t)$ for any time delay τ is given by

$$C_{xy}(\tau) = E[\{x(t) - \mu_x\}\{y(t + \tau) - \mu_y\}]$$
$$= \lim_{T \to \infty} \frac{1}{T} \int_0^T \{x(t) - \mu_x\}\{y(t + \tau) - \mu_y\} \, dt = R_{xy}(\tau) - \mu_x \mu_y \tag{3.15}$$

where

$$R_{xy}(\tau) = \lim_{T \to \infty} \frac{1}{T} \int_0^T x(t)y(t + \tau) \, dt \tag{3.16}$$

For the general case where $x(t)$ and $y(t)$ represent different data, $R_{xy}(\tau)$ in Equation 3.16 is called the *cross-correlation function* between $x(t)$ and $y(t)$. For the special case where $y(t) = x(t)$,

$$C_{xx}(\tau) = \lim_{T \to \infty} \frac{1}{T} \int_0^T \{x(t) - \mu_x\}\{x(t + \tau) - \mu_x\} \, dt = R_{xx}(\tau) - \mu_x^2 \tag{3.17}$$

where

$$R_{xx}(\tau) = \lim_{T \to \infty} \frac{1}{T} \int_0^T x(t)x(t + \tau)\, dt \qquad (3.18)$$

is called the *autocorrelation function* of $x(t)$. It should be mentioned that some textbooks use the term correlation function to refer to the quantity defined here as the covariance function in Equation 3.15. Since the two functions are interrelated by $R_{xy}(\tau) = C_{xy}(\tau) + \mu_x \mu_y$, it follows that $R_{xy}(\tau) = C_{xy}(\tau)$ if the mean of either measurement is zero.

From its definition, the autocorrelation function is always an even function of τ, that is,

$$R_{xx}(-\tau) = R_{xx}(\tau) \qquad (3.19)$$

The cross-correlation function, however, is neither an even nor odd function of τ, but satisfies the relation

$$R_{xy}(-\tau) = R_{yx}(\tau) \qquad (3.20)$$

From Equations 2.27 to 2.29, the value of the autocorrelation function at $\tau = 0$ is the mean square value of the data, that is,

$$R_{xx}(0) = \psi_x^2 = \sigma_x^2 + \mu_x^2 \qquad (3.21)$$

Assuming there are no deterministic components in the data other than a nonzero mean, the autocorrelation function collapses to a constant value equal to the square of the mean as τ increases, that is,

$$R_{xx}(\infty) = \mu_x^2 \qquad (3.22)$$

These various properties of autocorrelation and autocovariance functions are illustrated in Figure 3.2.

For the cross-correlation function, using a proof similar to that employed to arrive at Equation 3.4, it follows that

$$|R_{xy}(\tau)| \le \sqrt{R_{xx}(0)R_{yy}(0)} \qquad (3.23)$$

The relationship in Equation 3.23 is commonly called the *cross-correlation inequality*. The same relationship applies to the auto- and cross-covariance functions which gives rise to the *correlation coefficient function*.

$$\rho_{xy}(\tau) = \frac{C_{xy}(\tau)}{\sqrt{C_{xx}(0)C_{yy}(0)}} = \frac{R_{xy}(\tau) - \mu_x \mu_y}{\sqrt{[R_{xx}(0) - \mu_x^2][R_{yy}(0) - \mu_y^2]}} \qquad (3.24)$$

where $|\rho_{xy}(\tau)| \le 1$ for all τ. From Equation 3.20, the cross-correlation function is generally not an even function and $R_{xy}(0)$ has no specific relationship

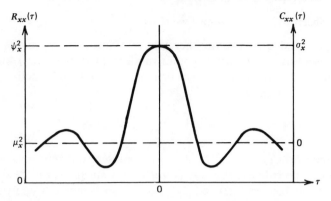

Figure 3.2 Properties of autocorrelation and autocovariance functions.

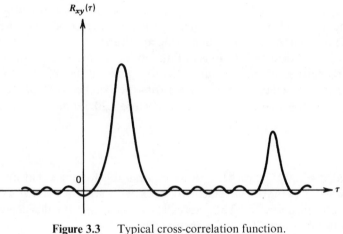

Figure 3.3 Typical cross-correlation function.

to the mean square values of the data records. A typical cross-correlation plot is shown in Figure 3.3.

3.2 SPECTRAL DENSITY FUNCTIONS

As noted earlier, the spectral density function can be thought of solely in terms of the Fourier transform of a correlation function. However, it can also be viewed in terms of generalized Fourier analysis or even more practically in terms of analog filtering operations.

3.2.1 *Spectra via Correlation Functions*

The spectral density function between two time history records $x(t)$ and $y(t)$ representing stationary (ergodic) random processes $\{x(t)\}$ and $\{y(t)\}$ may be defined as the Fourier transform of the correlation function between those records as follows:

$$S_{xy}(f) = \int_{-\infty}^{\infty} R_{xy}(\tau)e^{-j2\pi f\tau}\, d\tau \tag{3.25}$$

For the general case where $x(t)$ and $y(t)$ represent different data, $S_{xy}(f)$ in Equation 3.25 is called the *cross-spectral density function*, or more simply the *cross-spectrum* between $x(t)$ and $y(t)$. For the special case where $y(t) = x(t)$,

$$S_{xx}(f) = \int_{-\infty}^{\infty} R_{xx}(\tau)e^{-j2\pi f\tau}\, d\tau \tag{3.26}$$

where $S_{xx}(f)$ is called the *autospectral density function* or *autospectrum* of $x(t)$, or sometimes the *power spectral density function* because of its historic origin and use in communications engineering applications.

The spectral density functions in Equations 3.25 and 3.26 are defined over all frequencies, both positive and negative, and will be referred to as *two-sided* spectra to emphasize this fact. From the symmetry properties of correlation functions given in Equations 3.19 and 3.20, it follows that

$$S_{xx}(-f) = S_{xx}(f) \tag{3.27}$$

$$S_{xy}(-f) = S_{xy}^*(f) = S_{yx}(f) \tag{3.28}$$

Two-sided spectral density functions are desirable for analytical studies, but in practice it is more convenient to work with spectra defined over non-negative frequencies only. These are called *one-sided* spectral density functions and from Equations 3.25 and 3.26 are given by

$$
\begin{aligned}
G_{xy}(f) &= 2S_{xy}(f) = 2\int_{-\infty}^{\infty} R_{xy}(\tau)e^{-j2\pi f\tau}\, d\tau \quad & f \geq 0 \\
&= 0 & f < 0
\end{aligned}
\tag{3.29}
$$

$$
\begin{aligned}
G_{xx}(f) &= 2S_{xx}(f) = 2\int_{-\infty}^{\infty} R_{xx}(\tau)e^{-j2\pi f\tau}\, d\tau \quad & f \geq 0 \\
&= 0 & f < 0
\end{aligned}
\tag{3.30}
$$

The one-sided spectral density functions, denoted by $G(f)$, will be used throughout this book for applications problems, but the two-sided spectra, denoted by $S(f)$, will be used for analytical developments. The relationship between the two, as defined in Equation 3.30, is illustrated in Figure 3.4.

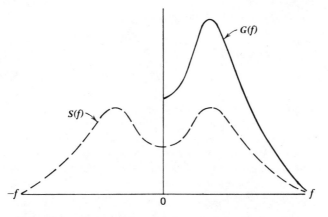

Figure 3.4   One- and two-sided spectral density functions.

Since autocorrelation functions are always even functions of τ, it follows that autospectra are given by the real part only of the Fourier transform in Equation 3.26. This yields

$$G_{xx}(f) = 2 \int_{-\infty}^{\infty} R_{xx}(\tau) \cos 2\pi f\tau \, d\tau = 4 \int_{0}^{\infty} R_{xx}(\tau) \cos 2\pi f\tau \, d\tau \quad (3.31)$$

Conversely, the autocorrelation function is given by the inverse transformation

$$R_{xx}(\tau) = \int_{-\infty}^{\infty} S_{xx}(f) e^{j2\pi f\tau} \, df = \int_{0}^{\infty} G_{xx}(f) \cos 2\pi f\tau \, df \quad (3.32)$$

From Equations 3.21 and 3.32, it is clear that at $\tau = 0$,

$$R_{xx}(0) = \int_{0}^{\infty} G_{xx}(f) \, df = \psi_x^2 = \sigma_x^2 + \mu_x^2 \quad (3.33)$$

In other words, the total area under the autospectral density function is the variance of the data plus the square of the mean value of the data. Furthermore, from Equation 3.17, since

$$R_{xx}(\tau) = C_{xx}(\tau) + \mu_x^2 \quad (3.34)$$

it follows from Equation 3.26 that

$$S_{xx}(f) = \int_{-\infty}^{\infty} C_{xx}(\tau) e^{-j2\pi f\tau} \, d\tau + \mu_x^2 \delta(f) \quad (3.35)$$

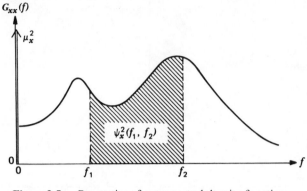

Figure 3.5 Properties of autospectral density functions.

where $\delta(f)$ is the delta function defined in Section 1.2.4. Hence a nonzero mean value appears in the spectral density function as a delta function at $f = 0$ with an area of μ_x^2. Finally, the area under the autospectrum between any two frequency limits f_1 and f_2 gives the mean square value of the data within that frequency range, that is,

$$\psi_x^2(f_1, f_2) = \int_{f_1}^{f_2} G_{xx}(f)\, df \tag{3.36}$$

These various properties of autospectra are illustrated in Figure 3.5.

The cross-correlation function $R_{xy}(\tau)$ is given by the inverse Fourier transform of the two-sided cross-spectrum $S_{xy}(f)$ of Equation 3.25. Hence

$$R_{xy}(\tau) = \int_{-\infty}^{\infty} S_{xy}(f) e^{j2\pi f \tau}\, df \tag{3.37}$$

In terms of the one-sided spectrum $G_{xy}(f)$, Equation 3.25 becomes

$$G_{xy}(f) = 2 \int_{-\infty}^{\infty} R_{xy}(\tau) e^{-j2\pi f \tau}\, d\tau = C_{xy}(f) - jQ_{xy}(f) \tag{3.38}$$

The real part

$$C_{xy}(f) = 2 \int_{-\infty}^{\infty} R_{xy}(\tau) \cos 2\pi f \tau\, d\tau \tag{3.39a}$$

is called the *coincident spectral density function*, or the *cospectrum* for short, and the imaginary part

$$Q_{xy}(f) = 2 \int_{-\infty}^{\infty} R_{xy}(\tau) \sin 2\pi f \tau\, d\tau \tag{3.39b}$$

is called the *quadrature spectral density function* or the *quadspectrum*. From Equation 3.37, in terms of $C_{xy}(f)$ and $Q_{xy}(f)$,

$$R_{xy}(\tau) = \frac{1}{2} \int_0^\infty G_{xy}(f)e^{j2\pi f\tau}\, df + \frac{1}{2} \int_0^\infty G_{xy}^*(f)e^{-j2\pi f\tau}\, df$$

$$= \int_0^\infty [C_{xy}(f)\cos 2\pi f\tau + Q_{xy}(f)\sin 2\pi f\tau]\, df \qquad (3.40)$$

Note that knowledge of $C_{xy}(f)$ can determine $R_{xy}(0)$. It is convenient and common in practice to present cross-spectra in terms of a *magnitude* and associated *phase angle* as follows:

$$G_{xy}(f) = |G_{xy}(f)|e^{-j\theta_{xy}(f)} \qquad (3.41)$$

where

$$|G_{xy}(f)| = \sqrt{C_{xy}^2(f) + Q_{xy}^2(f)} \qquad (3.41a)$$

$$\theta_{xy}(f) = \tan^{-1}\left[\frac{Q_{xy}(f)}{C_{xy}(f)}\right] \qquad (3.41b)$$

A typical cross-spectral density plot is shown in Figure 3.6.

The signs of the cross-spectral terms $C_{xy}(f)$ and $Q_{xy}(f)$ may be positive or negative and determine the quadrant for the phase angle $\theta_{xy}(f)$. These signs also determine at each frequency f whether $y(t)$ follows $x(t)$, as given by $y(t) = x(t - \tau_0)$ where $\tau_0 > 0$ represents a positive time delay for a signal to be transmitted from point x to point y. When signals at these two points are

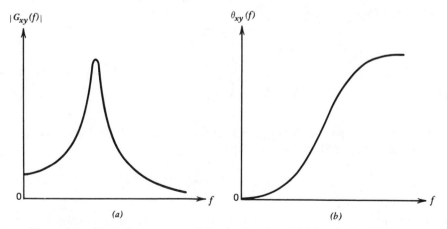

Figure 3.6   Typical cross-spectral density function. (*a*) Magnitude. (*b*) Phase.

measured with a common time base, observe that $y(t) = x(t - \tau_0)$ shows that $y(0)$ is due to $x(-\tau_0)$ and $y(\tau_0)$ is due to $x(0)$. A positive value of $\theta_{xy}(f)$ means that $y(t)$ follows $x(t)$ at frequency f, whereas a negative value of $\theta_{xy}(f)$ means that $x(t)$ follows $y(t)$ at frequency f.

An important relationship involving the magnitude of the cross-spectrum is given by the *cross-spectrum inequality*

$$|G_{xy}(f)|^2 \leq G_{xx}(f)G_{yy}(f) \tag{3.42}$$

This relationship, which is proved in the next section, is analogous to the cross-correlation inequality of Equation 3.23 and gives rise to the *coherence function*, defined as

$$\gamma_{xy}^2(f) = \frac{|G_{xy}(f)|^2}{G_{xx}(f)G_{yy}(f)} \qquad 0 \leq \gamma_{xy}^2(f) \leq 1 \tag{3.43}$$

which is analogous to the squared correlation coefficient function in Equation 3.24. However, the coherence function (sometimes called coherency squared) provides a far more powerful tool than the correlation coefficient function for many applications, as will be seen in later chapters.

3.2.2 Spectra via Fourier Transforms

A second way to develop spectral density functions is in terms of direct Fourier transformations of the original data records. Specifically, consider two stationary (ergodic) random processes $\{x(t)\}$ and $\{y(t)\}$, where the finite Fourier transforms over the kth record of length T representing each process are given by

$$X_k(f, T) = \int_0^T x_k(t)e^{-j2\pi ft} \, dt$$
$$\tag{3.44}$$
$$Y_k(f, T) = \int_0^T y_k(t)e^{-j2\pi ft} \, dt$$

The two-sided spectral density function between the two random processes is defined using X^*Y and *not* XY^* by

$$S_{xy}(f) = \lim_{T \to \infty} \frac{1}{T} E[X_k^*(f, T)Y_k(f, T)] \tag{3.45}$$

where the expected value operator E denotes an averaging operation over the index k. In terms of one-sided spectral density functions, the cross- and autospectra are given by

$$G_{xy}(f) = \lim_{T \to \infty} \frac{2}{T} E[X_k^*(f, T)Y_k(f, T)] \tag{3.46}$$

$$G_{xx}(f) = \lim_{T \to \infty} \frac{2}{T} E[|X_k(f, T)|^2] \tag{3.47}$$

The functions $S_{xy}(f)$, $S_{xx}(f)$, $G_{xy}(f)$, and $G_{xx}(f)$ defined in Equations 3.45 through 3.47 are identically equal to the corresponding functions defined in terms of Fourier transforms of correlation functions in Equations 3.25, 3.26, 3.29, and 3.30, respectively. This equality, commonly called the *Wiener-Khinchin relationship*, is proved in Reference 3.1. Hence all the relationships and properties of spectra developed in Section 3.2.1 apply here as well.

Using the spectral density function definition in Equation 3.45, the cross-spectrum inequality stated in Equation 3.42 can now be proved as follows: Given the finite Fourier transforms defined in Equation 3.44, for any real constants a and b and the cross-spectrum phase $\theta_{xy}(f)$, it follows that

$$\frac{1}{T} E[|aX_k(f, T) + bY_k(f, T)e^{j\theta_{xy}(f)}|^2] \geq 0 \tag{3.48}$$

Expanding the term in the brackets gives

$$\frac{1}{T} E[a^2|X_k(f, T)|^2 + abX_k^*(f, T)Y_k(f, T)e^{j\theta_{xy}(f)}$$

$$+ abX_k(f, T)Y_k^*(f, T)e^{-j\theta_{xy}(f)} + b^2|Y_k(f, T)|^2] \geq 0 \tag{3.49}$$

From Equation 3.45, taking the limit of Equation 3.49 as $T \to \infty$ yields

$$a^2 S_{xx}(f) + ab[S_{xy}(f)e^{j\theta_{xy}(f)} + S_{yx}(f)e^{-j\theta_{xy}(f)}] + b^2 S_{yy}(f) \geq 0 \tag{3.50}$$

However, from Equation 3.28, $S_{yx}(f) = S_{xy}^*(f)$, so

$$S_{xy}(f) = |S_{xy}(f)|e^{-j\theta_{xy}(f)} \qquad S_{yx}(f) = |S_{xy}(f)|e^{j\theta_{xy}(f)} \tag{3.51}$$

Substituting the relationships of Equation 3.51 into Equation 3.50 gives

$$a^2 S_{xx}(f) + 2ab|S_{xy}(f)| + b^2 S_{yy}(f) \geq 0 \tag{3.52}$$

Assuming $b \neq 0$, one can now divide through Equation 3.52 by b^2 to obtain

$$\left(\frac{a}{b}\right)^2 S_{xx}(f) + 2\left(\frac{a}{b}\right)|S_{xy}(f)| + S_{yy}(f) \geq 0 \tag{3.53}$$

which is a quadratic equation in (a/b) without real different roots, similar to Equation 3.7 in the proof of the cross-correlation inequality. As before, the

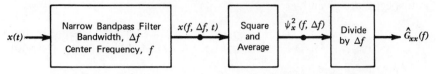

Figure 3.7   Autospectrum measurement by analog filtering operations.

discriminant of Equation 3.53 must be nonpositive, that is,

$$4|S_{xy}(f)|^2 - 4S_{xx}(f)S_{yy}(f) \leq 0 \qquad (3.54)$$

proving that

$$|S_{xy}(f)|^2 \leq S_{xx}(f)S_{yy}(f) \qquad (3.55)$$

From Equations 3.29 and 3.30, it follows that

$$|G_{xy}(f)|^2 \leq G_{xx}(f)G_{yy}(f)$$

thus proving Equations 3.42 and 3.43.

3.2.3   *Spectra via Analog Filtering*

Prior to the introduction and wide availability of digital signal processing equipment, spectral density functions, primarily autospectra, were commonly estimated using analog wave analyzers, as schematically illustrated in Figure 3.7. Many of these analog instruments are still in use. The principle of their operation is straightforward. A time history record $x(t)$ representing the data of interest is passed through a narrow bandpass filter of bandwidth Δf and variable center frequency f. The output of the filter, denoted by $x(f, \Delta f, t)$, is then squared, time averaged, and divided by Δf to obtain an autospectral density estimate as follows:

$$\hat{G}_{xx}(f) = \frac{1}{(\Delta f)T} \int_0^T x^2(f, \Delta f, t)\, dt \qquad (3.56)$$

If one takes the limit of the estimate in Equation 3.56 as $T \to \infty$ and $\Delta f \to 0$ such that $(\Delta f)T \to \infty$, the result will be the one-sided autospectral density function as previously defined in Equations 3.30 and 3.47, that is,

$$G_{xx}(f) = \lim_{\Delta f \to 0} \frac{1}{\Delta f} \left[\lim_{T \to \infty} \frac{1}{T} \int_0^T x^2(f, \Delta f, t)\, dt \right] \qquad (3.57)$$

The validity of this equality is proved in Reference 3.1.

3.3   GENERAL INTERPRETATIONS

The remaining chapters of this book deal directly with the applications of correlation and spectral density functions to various types of engineering problems. As an introduction to this material, a summary of the general interpretations of these functions may be helpful. Typical units for various quantities of interest are listed in Table 3.1.

3.3.1   *Autocorrelation Functions*

The autocorrelation function of random data is most directly interpreted as a measure of how well future values of the data can be predicted based on past observations. Of course, given the probability density function $p(x)$ of the data, one can always make statements concerning the probability that future time history values will fall within a specific interval, as discussed in Section 2.1.2. However, if the autocorrelation function at a given delay τ is different from μ^2 (or different from zero assuming $\mu = 0$), this suggests that a knowledge of the exact time history $x(t)$, $0 \leq t \leq T$, can help improve the prediction of $x(t)$ at the future time $t = T + \tau$, beyond the general probability statements provided by $p(x)$. To demonstrate this point, consider four common types of data that are often closely approximated in practice: (*a*) sine wave, (*b*) sine wave plus wide-band random noise, (*c*) narrow-band random noise, and (*d*) wide-band random noise. Typical time history records for these four cases are illustrated in Figure 3.8.

Consider first the sinusoidal data shown in Figure 3.8*a*. A sine wave can be viewed as the result of a stationary (ergodic) random process with sample records

$$x_k(t) = X \sin (2\pi f_0 t + \theta_k) \tag{3.58}$$

where θ_k is assumed to be uniformly distributed over 0 to 2π, that is,

$$p(\theta) = \frac{1}{2\pi} \qquad 0 \leq \theta \leq 2\pi \tag{3.59}$$

From Equation 3.15, the autocorrelation function of the sine wave is then given by

$$R_{xx}(\tau) = E[x_k(t)x_k(t + \tau)] \tag{3.60}$$

Substituting from Equation 3.58 and noting that the averaging is over θ as given by Equation 3.59, it follows that

$$R_{xx}(\tau) = \frac{X^2}{2\pi} \int_0^{2\pi} \sin (2\pi f_0 t + \theta) \sin [2\pi f_0(t + \tau) + \theta] \, d\theta \tag{3.61}$$

$$= \frac{X^2}{2} \cos 2\pi f_0 \tau$$

Table 3.1

Basic Quantities and Typical Units

Quantity	Units
$t, \Delta t, \tau$	sec (or other desired units)
$x(t), y(t)$	volts (or other desired units)
B, B_e, B_r	Hz
$C_{xx}(\tau), C_{xy}(\tau)$	(volts)^2
$f, \Delta f$	Hz
$G_{xx}(f), |G_{xy}(f)|$	(volts)^2 (sec) = (volts)^2/Hz
$R_{xx}(\tau), R_{xy}(\tau)$	(volts)^2
$S_{xx}(f), |S_{xy}(f)|$	(volts)^2 sec = (volts)^2/Hz
T, T_{total}	sec
$X(f, T), Y(f, T)$	(volts) (sec)
$|X(f, T)|^2, |Y(f, T)|^2$	(volts)^2 (sec)^2
$\gamma_{xy}^2(f), |\gamma_{xy}(f)|$	dimensionless
$\varepsilon_b, \varepsilon_r, \varepsilon$	dimensionless
$\theta_{xy}(f)$	radians
μ_x, μ_y	volts
$\rho_{xy}, \rho_{xy}(\tau)$	dimensionless
$\sigma_x^2, \sigma_y^2, \sigma_{xy}$	(volts)^2
ψ_x^2, ψ_y^2	(volts)^2

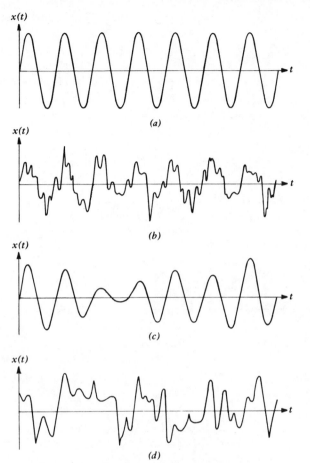

Figure 3.8 Four special time histories. (*a*) Sine wave. (*b*) Sine wave plus random noise. (*c*) Narrow-band random noise. (*d*) Wide-band random noise.

Hence the autocorrelation function of a sine wave is a cosine wave with an amplitude equal to the mean square value of the original sine wave, as illustrated in Figure 3.9*a*. The key observation here is that the envelope of the sine wave correlation function remains constant over all time delays, suggesting that one can predict future values of the data precisely based on past observations. Referring back to the time history record in Figure 3.8*a*, this is clearly correct, that is, having seen the time history of record length T, one can readily predict the exact value of the sine wave at any time in the future, assuming the data remain stationary.

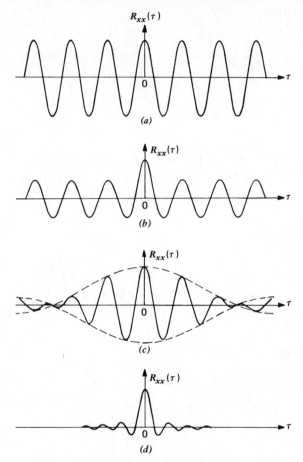

Figure 3.9 Idealized autocorrelation functions. (*a*) Sine wave. (*b*) Sine wave plus random noise. (*c*) Narrow-band random noise. (*d*) Wide-band random noise.

Next consider the wide-band random noise illustrated in Figure 3.8*d*. To establish an autocorrelation function for this case, assume the autospectrum of the data is uniform over a wide bandwidth B, that is,

$$G_{xx}(f) = G \qquad 0 \leq f \leq B$$
$$= 0 \qquad f > B \tag{3.62}$$

From Equation 3.32, the autocorrelation function is then given by

$$R_{xx}(\tau) = \int_0^B G \cos 2\pi f\tau \, df = GB\left(\frac{\sin 2\pi B\tau}{2\pi B\tau}\right) \tag{3.63}$$

This correlation function diminishes very rapidly with a first zero crossing at $\tau = 1/(2B)$, as illustrated in Figure 3.9d. The suggestion here is that a knowledge of the exact time history from zero to T will not significantly help one predict future values beyond the very near future, that is, beyond $1/(2B)$ past the end of the observed record. Referring again to the time history record in Figure 3.8d, this conclusion is clearly consistent with the visual character of the wide band random noise time history; that is, because of the eratic character of the time history, a past record does not significantly help one predict future values of the data beyond the very near future.

Concerning narrow-band random data, consider the ideal case where the autospectrum is uniform over a narrow bandwidth B centered at f_0, that is,

$$G_{xx} = G \qquad f_0 - \frac{B}{2} \leq f \leq f_0 + \frac{B}{2} \tag{3.64}$$

$$= 0 \qquad \text{otherwise}$$

Again from Equation 3.32, the autocorrelation function for this case is

$$R_{xx}(\tau) = GB\left(\frac{\sin \pi B\tau}{\pi B\tau}\right) \cos 2\pi f_0 \tau \tag{3.65}$$

This correlation function has an envelope that diminishes slowly, assuming B is small, with a first zero crossing at $\tau = 1/B$, as shown in Figure 3.9c. The suggestion here is that a knowledge of the exact time history from 0 to T will assist in predicting future values of $x(t)$ with diminishing precision. Referring again to the time history in Figure 3.8c, this conclusion is consistent with the almost single frequency character of the data. Specifically, because of the slowly varying envelope of the time history, the past observations can help improve the prediction of values in the near future, but since the envelope variations are random, not in the distant future.

The final case of interest is a sine wave plus wide-band random noise which might produce a time history as shown in Figure 3.8b. The autocorrelation function for this case is simply the sum of the autocorrelations of the sine wave and wide-band random noise, as given in Equations 3.61 and 3.63, respectively. That is,

$$R_{xx}(\tau) = \frac{X^2}{2} \cos 2\pi f_0 \tau + GB\left(\frac{\sin 2\pi B\tau}{2\pi B\tau}\right) \tag{3.66}$$

where X is the amplitude of the sine wave and G is the spectral density of the noise. A plot of a typical correlation function of this type is shown in Figure 3.9b. Note that the autocorrelation function decays quickly to the cosine term describing the sinusoidal portion of the data, suggesting that one can predict values into the distant future with imperfect but greater precision than provided by probability density information alone. Again referring to

Figure 3.8b, this conclusion is consistent with the visual character of the time history data.

Autocorrelation functions can also be interpreted in the context of dominant data frequencies, but information of this type is more easily extracted from autospectral density functions. However, the relationships between the autocorrelation function and the overall mean square value, mean value, and variance of the data, as given in Equations 3.21 and 3.22, are important to general interpretations and should be remembered. A final property of the autocorrelation function which is important to remember is the fact that it contains no phase information. Referring to the autocorrelation function derived for the case of the sine wave in Equation 3.61, the phase of the cosine function describing the sine wave autocorrelation is zero independent of the initial phase of the sine wave.

3.3.2 *Autospectral Density Functions*

Since the autospectrum is a frequency domain function, it is most directly interpreted as a measure of the frequency distribution of the mean square value of the data; that is, the rate of change of mean square value with frequency. The key relationship for general interpretations is given by Equation 3.36, namely

$$\psi_x^2(f_1, f_2) = \int_{f_1}^{f_2} G_{xx}(f)\, df \tag{3.67}$$

where $\psi_x^2(f_1, f_2)$ is the mean square value of the data within the frequency range f_1 to f_2. To illustrate this important property, again consider the four classes of data producing the typical time histories shown in Figure 3.8. The autospectra of these data are shown in Figure 3.10. These results can be compared directly to the corresponding autocorrelation functions in Figure 3.9.

Considering first the case of the sine wave, the autospectrum is given by Equations 3.31 and 3.61 as

$$G_{xx}(f) = 4 \int_0^\infty \frac{X^2}{2} \cos 2\pi f_0 \tau \, \cos 2\pi f\tau \, d\tau = \frac{X^2}{2} \delta(f - f_0) \tag{3.68}$$

where $\delta(f - f_0)$ is a delta function at $f = f_0$, as illustrated in Figure 3.10a. From the definition of the delta function discussed in Section 1.2.4,

$$G_{xx}(f) = \begin{cases} \infty & f = f_0 \\ 0 & f \neq f_0 \end{cases}$$

$$\int_{f_0 - \varepsilon}^{f_0 + \varepsilon} G_{xx}(f)\, df = \frac{X^2}{2} \tag{3.69}$$

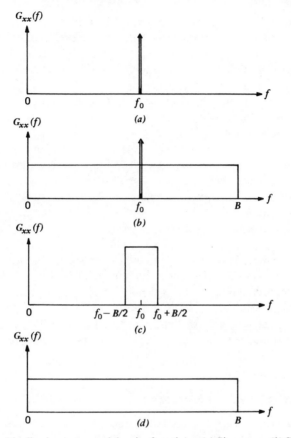

Figure 3.10 Idealized autospectral density functions. (*a*) Sine wave. (*b*) Sine wave plus random noise. (*c*) Narrow-band random noise. (*d*) Wide-band random noise.

where ε is any arbitrarily small value. The interpretation here is that the total mean square value of a sine wave, given by $\psi_x^2 = X^2/2$, is concentrated at the single frequency $f = f_0$. Such data generally suggest the existence of a cyclical physical phenomenon. Of course, the ideal result in Equation 3.68 will never be realized because there are no perfect sinusoidal phenomena that perpetuate over all time. Nevertheless, the spectral density function defined in Equation 3.68 often provides a good approximation to the data for many cyclical phenomena, for example, the vibration response of a rotating machine or the acoustic noise generated by a fan.

Now stepping ahead to the wide-band noise data, the autospectrum of this case, as previously defined in Equation 3.62, is illustrated in Figure 3.10*d*. The general interpretation of such data is obvious. For wide-band noise, the mean square value of the data is distributed over a wide frequency range, suggesting a broadly stochastic phenomenon, for example, turbulence data. The idealized case shown in Figure 3.10*d*, where the spectral density is uniform over the frequency range zero to *B*, is called *band-limited white noise*. In practice, random phenomena usually display some variations in their spectral density, but ideal band-limited white noise often provides a convenient assumption for certain theoretical studies and applications, to be discussed in later chapters.

For the case of the narrow-band noise shown in Figure 3.10*c*, the mean square value of the data is distributed over a narrow frequency range, but is not concentrated at a single frequency as for the sine wave case. The specific narrow-band spectrum shown in Figure 3.10*c* is defined in Equation 3.64 and represents an idealized case, but one which can sometimes be used as a first-order approximation in practice. More realistic narrow-band spectra typical of structural response data are developed in Chapter 5.

When a sine wave is mixed with wide-band noise, the resulting spectrum is the sum of the individual spectra, as shown in Figure 3.10*b*. The usual implication of such data is that a cyclical phenomenon is present in a broadly stochastic background. In practice, it is often difficult to distinquish between this situation and the narrow-band noise in Figure 3.10*c* due to the finite resolution associated with actual data analysis procedures, to be discussed in Section 3.4.2.

3.3.3   *Cross-Correlation Functions*

The most straightforward interpretations of cross-correlation functions are in the context of propagation problems. Specifically, assume that a physical phenomenon of interest, represented by a time history $x(t)$, propagates through a nondispersive linear path and is mixed with statistically independent noise $n(t)$ to produce a response measurement $y(t)$, as illustrated in Figure 3.11. For the simple case where the frequency response function of the propagation path is a constant $H(f) = H$, the propagation distance is d, and the propagation velocity is c, it follows that

$$y(t) = Hx\left(t - \frac{d}{c}\right) + n(t) \tag{3.70}$$

From Equations 3.16 and 3.18,

$$R_{xy}(\tau) = \lim_{T \to \infty} \frac{1}{T} \int_0^T x(t)\left[Hx\left(t - \frac{d}{c} + \tau\right) + n(t)\right] dt = HR_{xx}\left(\tau - \frac{d}{c}\right)$$

$$\tag{3.71}$$

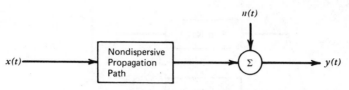

Figure 3.11   Nondispersive propagation model.

In other words, the cross-correlation function for this ideal case is given by the autocorrelation function of $x(t)$ multiplied by H and displaced in time to have a peak at $\tau_1 = (d/c)$. Hence given the knowledge of either the distance d or the velocity c, the other can be determined from the cross-correlation measurement. Furthermore, using the correlated output power relationship in Equation 3.10 along with the definition of the correlation coefficient function in Equation 3.24, the total variance of $y(t)$ that can be attributed to $x(t)$ alone is given by

$$\sigma_{y:x}^2 = \rho_{xy}^2\left(\frac{d}{c}\right)\sigma_y^2 \qquad (3.72)$$

where

$$\rho_{xy}^2\left(\frac{d}{c}\right) = \frac{[R_{xy}(d/c) - \mu_x\mu_y]^2}{[R_{xx}(0) - \mu_x^2][R_{yy}(0) - \mu_y^2]} = H^2\frac{\sigma_x^2}{\sigma_y^2}$$

Hence the fractional portion of $y(t)$ due to $x(t)$ in power terms can also be established from the cross-correlation measurement.

The above general interpretations of the cross-correlation function lead directly to applications involving ranging problems (determine d given c) and velocity measurement problems (determine c given d). Problems of this type are addressed in Chapter 7. Of particular interest are applications to multiple path problems where $x(t)$ can propagate through several possible paths to yield an output measurement $y(t)$. Assuming the propagation is wide-band and nondispersive (the velocity is not dependent on frequency), a cross-correlation measurement will theoretically yield a collection of correlation peaks, each defining the contribution through a given path, as illustrated in Figure 3.12. Applications of this type are detailed in Chapter 6. Cross-correlation functions can also be used to define the contribution of each of several independent sources of excitation to an output measurement, as illustrated in Figure 3.13, but problems of this type are usually approached with spectral analysis techniques.

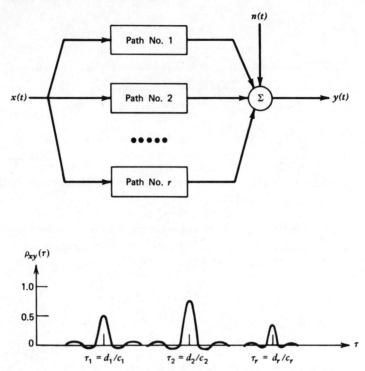

Figure 3.12   Typical cross-correlation results for nondispersive propagation through multiple paths.

3.3.4   *Cross-Spectral Density Functions*

The cross-spectral density function is generally interpreted like the cross-correlation function, except it provides the desired results as a function of frequency rather than in terms of overall values. This fact adds enormously to the scope of the interpretations and has lead in recent years to the increasing application of cross-spectra procedures to engineering problems where correlation analysis is relevent.

To illustrate the general interpretations of cross-spectral density functions, consider the simple nondispersive propagation problem shown previously in Figure 3.11. As will be proved later in Chapter 4, the cross-spectrum between the input $x(t)$ and the output $y(t)$, analogous to the cross-correlation function calculated in Equation 3.71, is

$$G_{xy}(f) = H(f)G_{xx}(f) = HG_{xx}(f)e^{-j2\pi f d/c} \tag{3.73}$$

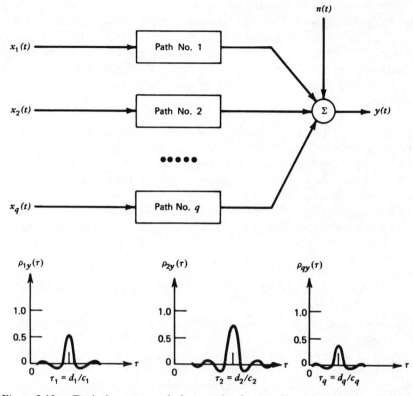

Figure 3.13 Typical cross-correlation results for nondispersive propagation from multiple sources.

Hence the propagation time $\tau_1 = (d/c)$ appears in the cross-spectrum as a linear phase shift given from Equation 3.41 by

$$\theta_{xy}(f) = 2\pi f \tau_1 = 2\pi f\left(\frac{d}{c}\right) \tag{3.74}$$

The important advantage of a cross-spectrum analysis over cross-correlation analysis is that the propagation need not be nondispersive to obtain meaningful results. Specifically, for a fixed distance d, the propagation velocity c is given as a function of frequency by

$$c(f) = \frac{2\pi f d}{\theta_{xy}(f)} \tag{3.75}$$

where $\theta_{xy}(f)$ is in radians and is not necessarily a linear function of frequency.

Correlation analysis has definite virtues over spectral analysis procedures for path identification problems as shown in Figure 3.12, to be discussed in Chapter 6. However, for the source identification problem illustrated in Figure 3.13, spectral analysis techniques are now used almost exclusively. Specifically, the analogy to the correlated output power relationship in Equation 3.72 is the coherent output power relationship given by

$$G_{y:x}(f) = \gamma_{xy}^2(f)G_{yy}(f) \tag{3.76}$$

By using coherence functions rather than correlation coefficient functions, the contribution of an input $x(t)$ to the measured output $y(t)$ is identified as a function of frequency rather than in overall terms only. Applications of this type are addressed in Chapter 9. Finally, spectral density functions provide a convenient vehicle to directly estimate properties of physical systems from input/output data which readily extend to multiple input systems. These matters are covered in Chapters 4 and 5, and for more advanced applications in Chapters 8 and 10.

3.4   ESTIMATION PROCEDURES AND ERRORS

The procedures for estimating correlation and spectral density functions with both analog instruments and digital computers are detailed in Reference 3.1. It is necessary here only to outline those aspects of the estimation procedures which relate to the error problem. All estimation procedures are discussed in the context of integral expressions which readily convert to sum expressions appropriate for digital data analysis procedures by noting the following facts which evolve directly from sampling considerations introduced in Section 1.2.3. First, given an analog time history record of length T, the record can be converted to a sequence of N equally spaced sample values with no significant loss of information if

$$N \geq 2BT \tag{3.77}$$

where B is the cyclical frequency bandwidth of the data (B is measured in Hz when T is in sec). Assuming the *time resolution* interval between the sample values is Δt, it follows that

$$T = N\Delta t \tag{3.78}$$

Next, the highest frequency that will appear in the sampled data, called the Nyquist cutoff or folding frequency, is given by

$$f_c = \frac{1}{2\Delta t} \tag{3.79}$$

where f_c is in Hz when Δt is in sec. To avoid the possiblity of serious errors due to aliasing effects, it is important that the data have no significant mean square value in the frequency range above f_c, as defined in Equation 3.79, prior to digital conversion. Finally, the minimum *frequency resolution* bandwidth available from the data record is given by

$$\Delta f = B_e = \frac{1}{T} = \frac{1}{N\Delta t} \tag{3.80}$$

where Δf is in Hz when T and Δt are in sec.

3.4.1   *Correlation Estimation Procedures*

From Equation 3.18, an unbiased estimate of the autocorrelation function of a stationary (ergodic) random process $\{x(t)\}$ based on a single sample record $x(t), 0 \le t \le T$, is given by

$$\hat{R}_{xx}(\tau) = \frac{1}{T - \tau} \int_0^{T-\tau} x(t)x(t + \tau)\, dt \tag{3.81}$$

Similarly, from Equation 3.16, an unbiased estimate of the cross-correlation function between two stationary (ergodic) processes $\{x(t)\}$ and $\{y(t)\}$ based on two sample records $x(t)$ and $y(t)$, both on a common time base $0 \le t \le T$, is given by

$$\hat{R}_{xy}(\tau) = \frac{1}{T - \tau} \int_0^{T-\tau} x(t)y(t + \tau)\, dt \tag{3.82}$$

Although Equations 3.81 and 3.82 provide the most direct method of estimating correlation functions, it is more efficient with modern digital signal processing equipment, and therefore more common today to estimate correlation functions by computing the finite inverse Fourier transform of spectral density estimates. Specifically, from Equations 3.32 and 3.40,

$$\hat{R}^c_{xx}(\tau) = \int_0^{f_c} \hat{G}_{xx}(f) \cos 2\pi f\tau \, df \tag{3.83}$$

$$\hat{R}^c_{xy}(\tau) = \int_0^{f_c} [\hat{C}_{xy}(f) \cos 2\pi f\tau + \hat{Q}_{xy}(f) \sin 2\pi f\tau]\, df \tag{3.84}$$

where $\hat{G}_{xx}(f)$, $\hat{C}_{xy}(f)$, and $\hat{Q}_{xy}(f)$ are "untapered" estimates of the auto, coincident, and quadrature spectral density functions, respectively, to be discussed in the next section.

The correlation estimates $\hat{R}^c(\tau)$ given by Equations 3.83 and 3.84 are subject to a bias error due to a circular effect in the calculation procedure, indicated by the superscript c. Specifically, the finite inverse Fourier transform operation is really an inverse Fourier series calculation that produces a circular correlation estimate $\hat{R}^c(\tau)$ instead of the desired $\hat{R}(\tau)$ of Equations 3.81 and 3.82, where

$$\hat{R}^c(\tau) = \frac{(T - \tau)}{T} \hat{R}(\tau) + \frac{\tau}{T} \hat{R}(T - \tau) \tag{3.85}$$

This is illustrated for an autocorrelation estimate in Figure 3.14.

The usual way to eliminate this circular effect in autocorrelation (with similar steps for cross-correlation) estimates is to extend the original time record $x(t)$, $0 \le t \le T$, by an additional record segment of length T with $x(t) = 0$ [or $x(t) = \hat{\mu}_x$ if the mean is nonzero]. In other words, assuming a zero mean value, a new record $w(t)$ of length $2T$ is created such that

$$
\begin{aligned}
w(t) &= x(t) & 0 \le t \le T \\
&= 0 & T < t \le 2T
\end{aligned}
\tag{3.86}
$$

When the spectrum of this longer record is computed and then inverse Fourier transformed, the resulting correlation function will now be

$$
\begin{aligned}
\hat{R}^s(\tau) &= \frac{(T - \tau)}{T} \hat{R}(\tau) & 0 \le \tau \le T \\
&= \frac{(\tau - T)}{T} \hat{R}(2T - \tau) & T < \tau \le 2T
\end{aligned}
\tag{3.87}
$$

That is, the desired autocorrelation function in Figure 3.14 and its mirror image will be separated as shown in Figure 3.15. One can now discard the

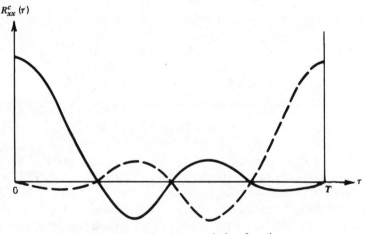

Figure 3.14   Circular correlation function.

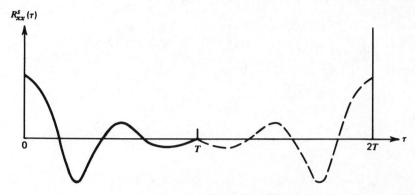

Figure 3.15   Circular correlation function for zero augmented record.

estimates for $\tau > T$ to obtain a correlation estimate with the circular effects removed over the range $0 \leq \tau \leq T$. However, a "bow-tie" correction is required to remove the ramp bias error in the correlation estimate as indicated by the coefficients of the $\hat{R}(\tau)$ terms in Equation 3.87. Thus an unbiased estimate of $\hat{R}(\tau)$ over the range $0 \leq \tau \leq T$ is given by

$$\hat{R}(\tau) = \frac{T}{T - \tau} \, \hat{R}^s(\tau) \tag{3.88}$$

The estimate $\hat{R}(\tau)$ in Equation 3.88 is statistically equivalent to the correlation estimates given by the direct calculation procedure in Equations 3.81 and 3.82.

3.4.2   *Spectral Estimation Procedures*

Spectral density functions can be estimated either through finite Fourier transforms of correlation functions, as suggested by Equations 3.29 and 3.30, or through finite Fourier transforms of the original time history records, as indicated by Equations 3.46 and 3.47. Since the introduction in 1965 of algorithms for the fast computation of Fourier series [3.2], the latter approach has become dominant. To execute this analysis approach in practice, the expected value operation in Equations 3.46 and 3.47 is accomplished by estimating spectral quantities for each of a collection of sample records and averaging the results. For stationary (ergodic) random data, this collection of records can be acquired sequentially in time as shown in Figure 3.16. Given a collection of n_d such records, $x_k(t)$; $(k - 1)T \leq t \leq kT$; $k = 1$,

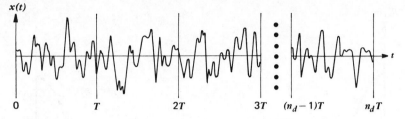

Figure 3.16   Subdivision of data into n_d records of individual length T.

$2, \ldots, n_d$, from a stationary (ergodic) random process $\{x(t)\}$, the autospectral density function is estimated from Equation 3.47 by

$$\hat{G}_{xx}(f) = \frac{2}{n_d T} \sum_{k=1}^{n_d} |X_k(f, T)|^2 \tag{3.89}$$

The total amount of data analyzed is $T_{\text{total}} = n_d T$. Similarly, for two collections of n_d paired records, $x_k(t)$ and $y_k(t)$; $(k-1)T \le t \le kT$; $k = 1, 2, \ldots, n_d$, from two stationary (ergodic) random processes $\{x(t)\}$ and $\{y(t)\}$, the cross-spectral density function is estimated from Equation 3.46 by

$$\hat{G}_{xy}(f) = \frac{2}{n_d T} \sum_{k=1}^{n_d} X_k^*(f, T) Y_k(f, T) \tag{3.90}$$

where $X_k(f, T)$ and $Y_k(f, T)$ in Equations 3.89 and 3.90 are the finite Fourier transforms of $x_k(t)$ and $y_k(t)$ defined previously by Equations 3.44 and 1.27.

It follows from Equation 3.43 that the coherence function is estimated by the algebraic operation of dividing $|\hat{G}_{xy}(f)|^2$ by the product of $\hat{G}_{xx}(f)$ with $\hat{G}_{yy}(f)$,

$$\hat{\gamma}_{xy}^2(f) = \frac{|\hat{G}_{xy}(f)|^2}{\hat{G}_{xx}(f)\hat{G}_{yy}(f)} \tag{3.91}$$

A meaningless result of $\hat{\gamma}_{xy}^2(f) = 1$ for all f occurs if $n_d = 1$. Note that the estimates given by Equations 3.89 through 3.91 appear as discrete frequency components separated in frequency by

$$B_e = \frac{1}{T} \tag{3.92}$$

where B_e constitutes the minimum resolution bandwidth of the analysis. It should be emphasized that this analysis resolution bandwidth is determined by the record length T and not by the total amount of data analyzed, $T_{\text{total}} = n_d T$.

The finite Fourier transform of $x(t)$ defined in Equation 3.44 can be viewed as the Fourier transform of an unlimited time history record $v(t)$ multiplied by a "rectangular" time window $u(t)$ where

$$u(t) = 1 \qquad 0 \le t \le T$$
$$ = 0 \qquad \text{otherwise}$$

(3.93)

In other words, the sample time history record $x(t)$ can be considered to be the product

$$x(t) = u(t)v(t)$$

(3.94)

It follows that the Fourier transform of $x(t)$ is the *convolution* of the Fourier transforms of $u(t)$ and $v(t)$, namely,

$$X(f) = \int_{-\infty}^{\infty} U(\alpha)V(f - \alpha)\, d\alpha$$

(3.95)

For the case where $u(t)$ is the "box car" function defined in Equation 3.93, its Fourier transform from Table 1.1 is

$$U(f) = T\left(\frac{\sin \pi f T}{\pi f T}\right)e^{-j\pi f T}$$

(3.96)

Plots of $u(t)$ and $|U(f)|$ are shown in Figure 3.17. $|U(f)|$ constitutes the basic "spectral window" of the analysis. The large side lobes of $|U(f)|$ allow leakage of power at frequencies well separated from the main lobe of the

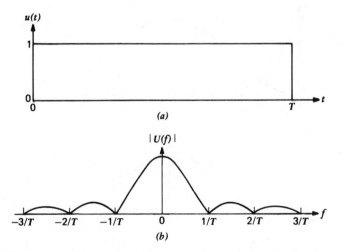

Figure 3.17   Rectangular analysis window. (*a*) Time window. (*b*) Spectral window.

spectral window and may introduce significant anomalies in the estimated spectra, particularly when the data are sinusoidal or narrow-band random in character. To suppress this problem, it is common in practice to introduce a time window that tapers the data so as to allow a more gradual entrance to and exit from the time history data to be analyzed. There are numerous such windows in current use, but one of the earliest and most commonly employed is a full cosine tapering window, called "Hanning," which is given by

$$u_h(t) = \frac{1}{2}\left(1 - \cos\frac{2\pi t}{T}\right) \qquad 0 \leq t \leq T$$

$$= 0 \qquad\qquad\qquad \text{otherwise}$$

(3.97)

The Fourier transform of Equation 3.97 is

$$U_h(f) = \tfrac{1}{2}U(f) - \tfrac{1}{4}U(f - f_1) - \tfrac{1}{4}U(f + f_1)$$

(3.98)

where $f_1 = (1/T)$. Note that

$$U(f - f_1) = -T\left[\frac{\sin \pi(f - f_1)T}{\pi(f - f_1)T}\right]e^{-j\pi fT}$$

$$U(f + f_1) = -T\left[\frac{\sin \pi(f + f_1)T}{\pi(f + f_1)T}\right]e^{-j\pi fT}$$

(3.99)

Plots of $u_h(t)$ and $|U_h(f)|$ are shown in Figure 3.18.

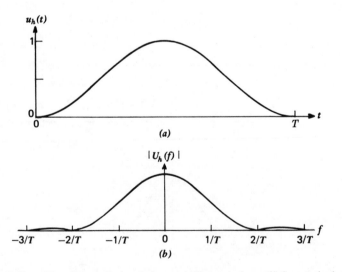

Figure 3.18   Hanning analysis window. (a) Time window. (b) Spectral window.

Consider now any function $v(t)$ that is *not* periodic of period T and let

$$x(t) = u_h(t)v(t) \qquad (3.100)$$

The Fourier transform of Equation 3.100 is

$$X(f) = \int_0^T x(t)e^{-j2\pi ft}\, dt \qquad (3.101)$$

At the discrete frequency values $f_k = (k/T)$ for $k = 0, 1, 2, \ldots, (N/2)$, one obtains

$$X(f_k) = \tfrac{1}{2}V(f_k) - \tfrac{1}{4}V(f_{k-1}) - \tfrac{1}{4}V(f_{k+1}) \qquad (3.102)$$

where

$$V(f_k) = \int_0^T v(t)e^{-j2\pi kt/T}\, dt \qquad (3.103)$$

To proceed further, assume that $v(t)$ behaves similar to band-limited white noise over the frequency resolution bandwidth $\Delta f = (1/T)$. It then follows that for any two discrete frequencies f and g calculated at the points $k\Delta f = (k/T)$, expected value operations on $V^*(f)$ and $V(g)$ will give

$$\begin{aligned} E[V^*(f)V(g)] &= 0 \qquad \text{for } f \neq g \\ &= 1 \qquad \text{for } f = g \end{aligned} \qquad (3.104)$$

Applying these properties to Equation 3.102 yields

$$E[|X(f_k)|^2] = (\tfrac{1}{2})^2 + (\tfrac{1}{4})^2 + (\tfrac{1}{4})^2 = \tfrac{3}{8} \qquad (3.105)$$

for any $f_k = (k/T)$, $k = 0, 1, 2, \ldots, (N/2)$. This represents a loss factor due to using the "Hanning" window of Equation 3.97 to compute spectral density estimates by finite Fourier transform techniques. Hence one should multiply Equation 3.102 by the scale factor $\sqrt{\tfrac{8}{3}}$ to obtain the correct magnitudes in later spectral density estimates using Equations 3.89 and 3.90. As a further observation, note that

$$\frac{\int_0^T u_h^2(t)\, dt}{\int_0^T u^2(t)\, dt} = \frac{3}{8} \qquad (3.106)$$

for the $u(t)$ and $u_h(t)$ of Equations 3.93 and 3.97.

In closing this brief summary of spectral analysis procedures, it should be mentioned that there are many other detailed operations and procedures that are commonly implemented in practice to enhance the quality of spectral density and coherence estimates. Included here are so-called "zoom" transform procedures that permit the analysis of data at high frequencies with very narrow resolution bandwidths, and "overlap" averaging procedures that retrieve some of the diminished statistical accuracy of the analyzed results due to side-lobe suppression (windowing) operations. See References 3.3, 3,4, and 3.5 for details.

3.4.3 *Statistical Sampling Errors*

The bias and random errors inherent in the computation of correlation and spectral density functions are derived in References 3.1, 3.6, and 3.7. A summary appears in Table 3.2. Statistical errors for more advanced functions are discussed in Chapter 11. The random error expressions for the correlation estimates in Table 3.2 constitute only guidelines since they assume the data have a constant spectrum over the entire frequency range of bandwidth B. The bias error term for a cross-spectrum estimate is an upper bound; if both records have a spectral peak at the same frequency, then the narrowest B_r of

Table 3.2

Summary of Correlation and Spectral Density Estimation Errors

Function Being Estimated	Estimation Procedure	Bias Error, ε_b	Random Error, ε_r
Auto or cross-correlation, $\hat{R}_{xx}(\tau)$ and $\hat{R}_{xy}(\tau)$	Equations 3.81 and 3.82	0	$\left[\dfrac{1 + \hat{\rho}_{xy}^{-2}(\tau)}{2BT_{\text{total}}}\right]^{1/2}$
	Equations 3.83 through 3.88	0	$\left[\dfrac{1 + \hat{\rho}_{xy}^{-2}(\tau)}{Nn_d}\right]^{1/2}$
Autospectrum, $\hat{G}_{xx}(f)$	Equation 3.89	$\dfrac{-1}{3}\left(\dfrac{B_e}{B_r}\right)^2$	$\dfrac{1}{\sqrt{n_d}}$
Cross-spectrum, $\|\hat{G}_{xy}(f)\|$	Equation 3.90	$\dfrac{-1}{3}\left(\dfrac{B_e}{B_r}\right)^2$	$\dfrac{1}{\|\hat{\gamma}_{xy}(f)\|\sqrt{n_d}}$
Coherence, $\hat{\gamma}_{xy}^2(f)$	Equation 3.91	Undefined	$\dfrac{\sqrt{2}[1 - \hat{\gamma}_{xy}^2(f)]}{\|\hat{\gamma}_{xy}(f)\|\sqrt{n_d}}$

B = Bandwidth of the data (assumed to be band-limited white noise).
B_e = Spectral resolution bandwidth defined in Equation 3.92.
B_r = Half-power point bandwidth of spectral peak defined in Equation 1.58.
n_d = Number of records illustrated in Figure 3.16.
N = Number of data points per record.

the two records should be used. Finally, the bias error in coherence estimates due to resolution problems can be quite complicated and surprisingly large, as will be illustrated later in Section 5.2.3. Reference 3.3 shows that there are also bias errors in coherence estimates due to sampling considerations, but they are generally negligible compared to the random errors.

REFERENCES

3.1 Bendat, J. S., and Piersol, A. G., *Random Data: Analysis and Measurement Procedures*, Wiley-Interscience, New York, 1971.

3.2 Cooley, J. W., and Tukey, J. W., "An Algorithm for the Machine Calculation of Complex Fourier Series," *Mathematics of Computation*, Vol. 19, No. 90, p. 297, April 1965.

3.3 Carter, G. C., Knapp, C. H., and Nuttall, A. H., "Estimation of the Magnitude-Squared Coherence via Overlapped Fast Fourier Transform Processing," *IEEE Transactions on Audio and Electroacoustics*, Vol. AU-21, No. 4, p, 337, August 1973.

3.4 Enochson, L. D., "Digital Techniques in Data Analysis," *Noise Control Engineering*, Vol. 9, No. 2, p. 138, November-December 1977.

3.5 Otnes, R. K., and Enochson, L. D., *Applied Time Series Analysis*, Wiley-Interscience, New York, 1978.

3.6 Bendat, J. S., *Principles and Applications of Random Noise Theory*, Wiley, New York, 1958. Reprinted by Krieger Publishing Co., New York, 1977.

3.7 Bendat, J. S., "Statistical Errors in Measurement of Coherence Functions and Input/Output Quantities," *Journal of Sound and Vibration*, Vol. 59, No. 3, p. 405, 1978.

CHAPTER 4

SINGLE INPUT/SINGLE OUTPUT RELATIONSHIPS

Evaluations of input/output data are fundamental to a wide class of engineering problems involving correlation and spectral analysis. This chapter develops the basic relationships for single input/single output problems. It is assumed here that the inputs are records from stationary (ergodic) random or transient random processes with zero mean values, and that systems are constant parameter linear systems, as defined in Chapter 1. Appropriate relationships for multiple input and multiple output problems are developed later in Chapters 7, 8, and 10.

4.1 IDEAL SYSTEM RELATIONSHIPS

Consider a constant parameter linear system with a weighting function $h(\tau)$ and frequency response function $H(f)$ as defined and discussed in Section 1.3. Assume that the system is subjected to a well-defined single input $x(t)$ which produces a well-defined output $y(t)$, as illustrated in Figure 4.1. Under ideal conditions as defined in Section 1.3.1, the output of the system in Figure 4.1 is given by the convolution integral

$$y(t) = \int_0^\infty h(\tau)x(t - \tau) \, d\tau \tag{4.1}$$

where $h(\tau) = 0$ for $\tau < 0$ when the system is physically realizable.

78

4.1.1   *Stationary Inputs*

Assume that the input $x(t)$ to the system in Figure 4.1 is a sample record from a stationary (ergodic) random process $\{x(t)\}$ as defined in Section 1.1. After switch-on transients have decayed, the response $y(t)$ will also belong to a stationary process $\{y(t)\}$. From Equation 4.1, the product $y(t)y(t + \tau)$ is given by

$$y(t)y(t + \tau) = \int_0^\infty \int_0^\infty h(\xi)h(\eta)x(t - \xi)x(t + \tau - \eta) \, d\xi \, d\eta \qquad (4.2)$$

From Equation 3.15, taking the expected values of both sides of Equation 4.2 yields the *input/output autocorrelation relation*

$$R_{yy}(\tau) = \int_0^\infty \int_0^\infty h(\xi)h(\eta)R_{xx}(\tau + \xi - \eta) \, d\xi \, d\eta \qquad (4.3)$$

Similarly, the product $x(t)y(t + \tau)$ is given by

$$x(t)y(t + \tau) = \int_0^\infty h(\xi)x(t)x(t + \tau - \xi) \, d\xi \qquad (4.4)$$

Here, expected values of both sides yield the *input/output cross-correlation relation*

$$R_{xy}(\tau) = \int_0^\infty h(\xi)R_{xx}(\tau - \xi) \, d\xi \qquad (4.5)$$

Note that Equation 4.5 is a convolution integral of the same form as Equation 4.1.

From Section 3.2.1, direct Fourier transforms of Equations 4.3 and 4.5 after various algebraic steps yield two-sided spectral density functions $S_{xx}(f)$, $S_{yy}(f)$, and $S_{xy}(f)$, which satisfy the important formulas

$$S_{yy}(f) = |H(f)|^2 S_{xx}(f) \qquad (4.6)$$

$$S_{xy}(f) = H(f)S_{xx}(f) \qquad (4.7)$$

Here, f may be either positive or negative. Note that Equation 4.6 is a real-valued relation containing only the gain factor $|H(f)|$ of the system. However, Equation 4.7 is a complex-valued relation that can be broken down into a pair of equations to give both the gain factor $|H(f)|$ and the phase factor

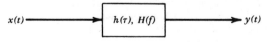

Figure 4.1   Ideal single input/single output system.

$\phi(f)$ of the system. Equation 4.6 is called the *input/output autospectrum relation*, whereas Equation 4.7 is called the *input/output cross-spectrum relation*. These results apply only to ideal situations where no extraneous noise exists at input or output points, and the systems have no time-varying or nonlinear characteristics. Interpretation of these spectral relations in the frequency domain is much easier than their corresponding correlation relations in the time domain [4.1].

In terms of physically measurable one-sided spectral density functions as defined in Section 3.2.1, Equations 4.6 and 4.7 become

$$G_{yy}(f) = |H(f)|^2 G_{xx}(f) \tag{4.8}$$

$$G_{xy}(f) = H(f)G_{xx}(f) \tag{4.9}$$

Using complex polar notation such that

$$\begin{aligned} G_{xy}(f) &= |G_{xy}(f)|e^{-j\theta_{xy}(f)} \\ H(f) &= |H(f)|e^{-j\phi(f)} \end{aligned} \tag{4.10}$$

then Equation 4.9 is equivalent to

$$\begin{aligned} |G_{xy}(f)| &= |H(f)|G_{xx}(f) \\ \theta_{xy}(f) &= \phi(f) \end{aligned} \tag{4.11}$$

From Equation 4.8, the output mean square value ψ_y^2 is given by

$$\psi_y^2 = \int_0^\infty G_{yy}(f) \, df = \int_0^\infty |H(f)|^2 G_{xx}(f) \, df \tag{4.12}$$

Equation 4.8 also permits the determination of $G_{xx}(f)$ from a knowledge of $G_{yy}(f)$ and $|H(f)|$, or the determination of $|H(f)|$ from a knowledge of $G_{xx}(f)$ and $G_{yy}(f)$. However, Equation 4.8 does not yield the complete frequency response function $H(f)$ of the system since it contains no phase information. The complete frequency response function in both gain and phase can only be obtained from Equations 4.9 to 4.11 when both $G_{xy}(f)$ and $G_{xx}(f)$ are known, as will be detailed later in Chapter 5.

An alternate and direct way is available to derive Equations 4.8 and 4.9 without first computing the correlation expressions of Equations 4.3 and 4.5. For any pair of truncated records of long length T, Equation 4.1 is equivalent to

$$Y(f, T) = H(f)X(f, T) \tag{4.13}$$

where $X(f, T)$ and $Y(f, T)$ are finite Fourier transforms of $x(t)$ and $y(t)$, respectively, as defined previously in Equation 3.44. It follows that

$$Y^*(f, T) = H^*(f)X^*(f, T)$$

$$|Y(f, T)|^2 = |H(f)|^2 |X(f, T)|^2$$

$$X^*(f, T)Y(f, T) = H(f)|X(f, T)|^2$$

From Equations 3.46 and 3.47, taking the expectation over different possible records, multiplying by $(2/T)$, and letting T increase beyond bound now proves the relations in Equations 4.8 and 4.9, namely

$$G_{yy}(f) = |H(f)|^2 G_{xx}(f)$$

$$G_{xy}(f) = H(f)G_{xx}(f)$$

4.1.2   Transient Inputs

For transient random records of length T, "energy" spectral density functions $\mathscr{G}_{xx}(f)$, $\mathscr{G}_{yy}(f)$, and $\mathscr{G}_{xy}(f)$ should be used instead of "power" spectral density functions $G_{xx}(f)$, $G_{yy}(f)$, and $G_{xy}(f)$. These are related by $\mathscr{G}_{xy}(f) = TG_{xy}(f)$ where the transient records $x(t)$ and $y(t)$ are assumed to exist only in the range $0 \le t \le T$. In place of Equations 4.8 and 4.9, one obtains

$$\mathscr{G}_{yy}(f) = |H(f)|^2 \mathscr{G}_{xx}(f)$$
$$\mathscr{G}_{xy}(f) = H(f)\mathscr{G}_{xx}(f)$$

$$(4.14a)$$

where f is restricted to be non-negative. Two-sided functions occur by setting $\mathscr{S}_{xy}(f) = (\frac{1}{2})\mathscr{G}_{xy}(f)$ for $f \ge 0$ and $\mathscr{S}_{xy}(-f) = \mathscr{S}^*_{xy}(f)$. When $x(t) = y(t)$, this gives $\mathscr{S}_{xx}(f) = (\frac{1}{2})\mathscr{G}_{xx}(f)$ for $f \ge 0$ and $\mathscr{S}_{xx}(-f) = \mathscr{S}_{xx}(f)$. Thus

$$\mathscr{S}_{yy}(f) = |H(f)|^2 \mathscr{S}_{xx}(f)$$
$$\mathscr{S}_{xy}(f) = H(f)\mathscr{S}_{xx}(f)$$

$$(4.14b)$$

Note that except for scale factors, results are exactly the same for stationary random or transient random records using one-sided or two-sided functions.

4.1.3   Effects of Nonzero Mean Values

Suppose the input stationary random process $\{x(t)\}$ has a nonzero mean value μ_x. From Equation 4.1, the output process $\{y(t)\}$ will have a nonzero mean value μ_y given by

$$\mu_y = \int_0^\infty h(\tau)\mu_x \, d\tau = H(0)\mu_x \qquad (4.15)$$

where $H(0)$ is the value of $H(f)$ as f approaches zero from the right. If μ_x and μ_y are measured using truncated records of length T, then their finite Fourier transforms (F.T.) can be interpreted as

$$\text{F.T.}[\mu_x] = \mu_x\delta_1(f) \qquad \text{F.T.}[\mu_y] = \mu_y\delta_1(f) \tag{4.16}$$

where the $\delta_1(f)$ is a *finite delta function* defined by

$$\delta_1(f) = T \quad \text{for} \left(\frac{-1}{2T}\right) \leq f \leq \left(\frac{1}{2T}\right) \tag{4.17}$$

$$= 0 \qquad \text{otherwise}$$

The associated two-sided autospectra can be represented by

$$S_{xx}(f) = \frac{1}{T} E[|\mu_x\delta_1(f)|^2] = \mu_x^2\delta_1(f)$$

$$\tag{4.18}$$

$$S_{yy}(f) = \frac{1}{T} E[|\mu_y\delta_1(f)|^2] = \mu_y^2\delta_1(f)$$

where $\delta_1(f)$ satisfies Equation 4.17. It follows that

$$\int_{-\infty}^{\infty} S_{xx}(f)\, df = \int_{-1/2T}^{1/2T} \mu_x^2\delta_1(f)\, df = \mu_x^2$$

$$\tag{4.19}$$

$$\int_{-\infty}^{\infty} S_{yy}(f)\, df = \int_{-1/2T}^{1/2T} \mu_y^2\delta_1(f)\, df = \mu_y^2$$

Corresponding one-sided autospectra should be defined here by

$$G_{xx}(f) = 2\mu_x^2\delta_1(f) \qquad \text{for } f \geq 0$$
$$G_{yy}(f) = 2\mu_y^2\delta_1(f) \qquad \text{for } f \geq 0 \tag{4.20}$$

so that the integral of $G_{xx}(f)$ gives an area of μ_x^2, that is,

$$\int_{0}^{\infty} G_{xx}(f)\, df = \int_{0}^{1/2T} 2\mu_x^2\delta_1(f)\, df = \mu_x^2$$

$$\tag{4.21}$$

$$\int_{0}^{\infty} G_{yy}(f)\, df = \int_{0}^{1/2T} 2\mu_y^2\delta_1(f)\, df = \mu_y^2$$

Any arbitrary input record $x(t)$ of length T can be considered to be the sum of two terms $[x(t) - \mu_x]$ and μ_x. In place of Equation 4.13, one will obtain

$$Y(f, T) + \mu_y\delta_1(f) = H(f)X(f, T) + H(0)\mu_x\delta_1(f) \tag{4.22}$$

where $X(f, T)$ and $Y(f, T)$ are the finite Fourier transforms of $[x(t) - \mu_x]$ and $[y(t) - \mu_y]$, respectively, and $\delta_1(f)$ satisfies Equation 4.17. Proceeding as before, in place of Equations 4.8 and 4.9, one will derive

$$G_{yy}(f) + 2\mu_y^2\delta_1(f) = |H(f)|^2 G_{xx}(f) + 2H^2(0)\mu_x^2\delta_1(f)$$
$$G_{xy}(f) = H(f)G_{xx}(f) \tag{4.23}$$

Note that no change occurs in the cross-spectrum relation of Equation 4.9, whereas the autospectrum relation of Equation 4.8 is affected only at points where $\delta_1(f) \neq 0$. In particular, if f is computed at discrete points that are spaced $B_e = (1/T)$ apart, then $\delta_1(f) = 0$ at all points except for the point $f = 0$ where $\delta_1(0) = T$.

With respect to Equation 4.12, the effect of a nonzero mean value on the total output mean square value is

$$\psi_y^2 = \int_0^\infty |H(f)|^2 G_{xx}(f)\,df + \mu_y^2 \tag{4.24}$$

where μ_y^2 is given by $\mu_y^2 = H^2(0)\mu_x^2$. These results show that nonzero mean values can be handled easily and separately by merely adding on appropriate terms. Measured data should be transformed, if required, so as to have non-zero mean values in all the succeeding discussions.

4.1.4 *Ordinary Coherence Functions*

The ordinary coherence function between $x(t)$ and $y(t)$ is a real-valued quantity defined earlier in Equation 3.43 by

$$\gamma_{xy}^2(f) = \frac{|G_{xy}(f)|^2}{G_{xx}(f)G_{yy}(f)} = \frac{|S_{xy}(f)|^2}{S_{xx}(f)S_{yy}(f)} \tag{4.25}$$

where the G's are the one-sided spectra and the S's are the corresponding two-sided spectra. It is proved in Section 3.2.2 that

$$0 \le \gamma_{xy}^2(f) \le 1 \tag{4.26}$$

Mean values different from zero should be removed from the data before computing $\gamma_{xy}^2(f)$ to eliminate delta function behavior at the origin. It is, of course, also required that both $G_{xx}(f)$ and $G_{yy}(f)$ should be greater than zero to avoid division by zero, or to have a zero over zero condition when $G_{xy}(f)$ is also close to zero.

For the ideal case satisfied by Equations 4.6 through 4.9, one obtains for all f

$$\gamma_{xy}^2(f) = \frac{|H(f)G_{xx}(f)|^2}{G_{xx}(f)|H(f)|^2 G_{xx}(f)} = 1 \tag{4.27}$$

On the other hand, if $x(t)$ and $y(t)$ are completely uncorrelated so that $G_{xy}(f) = 0$ for all f, then the coherence function $\gamma^2_{xy}(f) = 0$ for all f.

In actual practice, when the coherence function is greater than zero but less than unity, one or more of the following four main conditions exist:

1. Extraneous noise is present in the measurements.
2. Resolution bias errors are present in the spectral estimates.
3. The system relating $y(t)$ to $x(t)$ is not linear.
4. The output $y(t)$ is due to other inputs besides $x(t)$.

When extraneous noise $n(t)$ is present only at the output point, the total measured output spectrum $G_{yy}(f)$ consists of the sum of the ideal linear output $G_{vv}(f)$, due to $x(t)$ passing through $H(f)$, plus the output noise $G_{nn}(f)$. Specifically,

$$G_{yy}(f) = G_{vv}(f) + G_{nn}(f) \tag{4.28}$$

where

$$G_{vv}(f) = |H(f)|^2 G_{xx}(f) \tag{4.29}$$

It follows that

$$G_{vv}(f) = \left| \frac{G_{xy}(f)}{G_{xx}(f)} \right|^2 G_{xx}(f) = \gamma^2_{xy}(f) G_{yy}(f) \tag{4.30}$$

Hence

$$\gamma^2_{xy}(f) = \frac{G_{vv}(f)}{G_{yy}(f)} \tag{4.31}$$

The product of $\gamma^2_{xy}(f)$ with $G_{yy}(f)$ in Equation 4.30 is called the *coherent output (power) spectrum*. Equation 4.31 shows that the coherence function can be interpreted as the fractional portion of the output spectrum of $y(t)$ which is linearly due to $x(t)$ at frequency f. The *output noise spectrum* $G_{nn}(f) = G_{yy}(f) - G_{vv}(f)$ becomes

$$G_{nn}(f) = [1 - \gamma^2_{xy}(f)] G_{yy}(f) \tag{4.32}$$

and can be interpreted as the portion of the output spectrum which is not accounted for by linear operations on $x(t)$ at frequency f.

Substitution of Equation 4.28 into 4.31 also proves

$$\gamma^2_{xy}(f) = \frac{G_{yy}(f) - G_{nn}(f)}{G_{yy}(f)} = 1 - \frac{G_{nn}(f)}{G_{yy}(f)} \tag{4.33}$$

Thus the coherence function is a direct measure of the ratio of $G_{nn}(f)$ to the total measured $G_{yy}(f)$ at frequency f. Observe that $\gamma^2_{xy}(f)$ approaches unity as $G_{nn}(f)/G_{yy}(f)$ approaches zero, and conversely, $\gamma_{xy}^2(f)$ approaches zero as $G_{nn}(f)/G_{yy}(f)$ approaches unity. In terms of the output noise-to-signal ratio of $G_{nn}(f)$ to $G_{vv}(f)$, Equations 4.31 and 4.32 yield

$$\frac{G_{nn}(f)}{G_{vv}(f)} = \frac{1 - \gamma_{xy}^2(f)}{\gamma_{xy}^2(f)} \tag{4.34}$$

Here, $G_{nn}(f)/G_{vv}(f)$ approaches infinity as $\gamma_{xy}^2(f)$ approaches zero, and approaches zero as $\gamma_{xy}^2(f)$ approaches unity.

When extraneous noise $m(t)$ is present only at the input point, then the measured input record $x(t) = u(t) + m(t)$ where $u(t)$ is the true input signal. The measured input spectrum $G_{xx}(f)$ now consists of the sum of $G_{uu}(f)$ plus $G_{mm}(f)$, assuming $u(t)$ and $m(t)$ are uncorrelated. Specifically,

$$G_{xx}(f) = G_{uu}(f) + G_{mm}(f) \tag{4.35}$$

but now Equations 4.8 and 4.9 become

$$G_{yy}(f) = |H(f)|^2 G_{uu}(f)$$
$$G_{xy}(f) = H(f)G_{uu}(f) \tag{4.36}$$

It follows here that

$$\gamma_{xy}^2(f)G_{xx}(f) = \frac{|G_{xy}(f)|^2}{G_{yy}(f)} = G_{uu}(f) \tag{4.37}$$

Thus

$$\gamma_{xy}^2(f) = \frac{G_{uu}(f)}{G_{xx}(f)} = 1 - \frac{G_{mm}(f)}{G_{xx}(f)} \tag{4.38}$$

An important useful property of coherence functions is that they are preserved under linear transformations. Thus suppose $\gamma_{xy}^2(f)$ is a desired coherence function between $x(t)$ and $y(t)$. Now if $x_1(t)$ is perfectly linearly related to $x(t)$ and $y_1(t)$ is perfectly linearly related to $y(t)$, then $\gamma_{x_1y_1}^2(f) = \gamma_{xy}^2(f)$. This result is helpful in physical problems where it may be convenient to measure $x_1(t)$ and/or $y_1(t)$ instead of $x(t)$ and/or $y(t)$ to obtain a desired $\gamma_{xy}^2(f)$, as will arise later in Chapter 9.

4.2   EFFECTS OF MEASUREMENT NOISE

Consider a more realistic model where extraneous measurement noise exists at both input and output points. Let the true signals be $u(t)$ and $v(t)$, and the noise terms be $m(t)$ and $n(t)$, respectively, as shown in Figure 4.2.

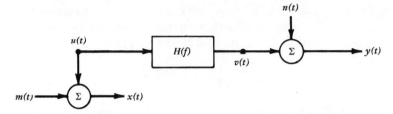

Figure 4.2   Single input/single output system with extraneous noise.

4.2.1   *Uncorrelated Input and Output Noise*

With noise at both the input and output points, it follows that the measured input and output records will be

$$x(t) = u(t) + m(t)$$
$$y(t) = v(t) + n(t) \tag{4.39}$$

where it will be assumed that the cross-spectral terms

$$G_{xm}(f) = G_{yn}(f) = G_{mn}(f) = 0 \tag{4.40}$$

Thus the extraneous noise terms are uncorrelated with each other and with the signals. Cases where $G_{mn}(f) \neq 0$ are treated in Section 4.2.3. In terms of true $G_{uu}(f)$, $G_{vv}(f)$, and $G_{uv}(f)$, Equations 4.8 and 4.9 give

$$G_{vv}(f) = |H(f)|^2 G_{uu}(f)$$
$$G_{uv}(f) = H(f)G_{uu}(f) \tag{4.41}$$

However, the measured spectral density functions are $G_{xx}(f)$, $G_{yy}(f)$, and $G_{xy}(f)$, where

$$G_{xx}(f) = G_{uu}(f) + G_{mm}(f) \geq G_{uu}(f)$$
$$G_{yy}(f) = G_{vv}(f) + G_{nn}(f) \geq G_{vv}(f) \tag{4.42}$$

$$G_{xy}(f) = G_{uv}(f) \tag{4.43}$$

since $G_{mm}(f) \geq 0$ and $G_{nn}(f) \geq 0$ for all f.
   Now, the true coherence function is given by

$$\gamma_{uv}^2(f) = \frac{|G_{uv}(f)|^2}{G_{uu}(f)G_{vv}(f)} \tag{4.44}$$

However, the measured coherence function is

$$\gamma_{xy}^2(f) = \frac{|G_{xy}(f)|^2}{G_{xx}(f)G_{yy}(f)} \tag{4.45}$$

From Equations 4.41 through 4.43,

$$|G_{xy}(f)| = |G_{uv}(f)|^2 = |H(f)|^2 G_{uu}^2(f) = G_{uu}(f)G_{vv}(f) \qquad (4.46)$$

Hence the coherence function becomes

$$\gamma_{xy}^2(f) = \frac{G_{uu}(f)G_{vv}(f)}{[G_{uu}(f) + G_{mm}(f)][G_{vv}(f) + G_{nn}(f)]} \leq 1 \qquad (4.47)$$

Strict inequality results whenever $G_{mm}(f) > 0$ or $G_{nn}(f) > 0$, which will always be the case with real data. It now follows from the definition of the coherent output power spectrum in Equation 4.30 that

$$\gamma_{xy}^2(f)G_{yy}(f) = G_{vv}(f)\left[\frac{G_{uu}(f)}{G_{uu}(f) + G_{mm}(f)}\right] \qquad (4.48)$$

Thus the coherent output power spectrum will determine $G_{vv}(f)$ when $G_{mm}(f) = 0$, regardless of the output noise $G_{nn}(f)$.

By the autospectrum method of Equation 4.8, one way to estimate the square of the system gain factor is by

$$|H(f)|_a^2 = \frac{G_{yy}(f)}{G_{xx}(f)} \qquad (4.49)$$

By the cross-spectrum method of Equation 4.9, a second way to estimate the square of the system gain factor is by

$$|H(f)|_c^2 = \frac{|G_{xy}(f)|^2}{G_{xx}^2(f)} \qquad (4.50)$$

The ratio of these two equations gives the coherence function by the formula

$$\gamma_{xy}^2(f) = \frac{|H(f)|_c^2}{|H(f)|_a^2} = \frac{|G_{xy}(f)|^2}{G_{xx}(f)G_{yy}(f)} \qquad (4.51)$$

In practice, since $\gamma_{xy}^2(f) < 1$, the gain factor estimate by the cross-spectrum method will always give a lower estimate than the gain factor estimate by the autospectrum method. Furthermore, substitution of Equation 4.42 into Equation 4.49 shows that

•

$$|H|_a^2 = \frac{G_{vv} + G_{nn}}{G_{uu} + G_{mm}} = |H|^2\left[\frac{1 + (G_{nn}/G_{vv})}{1 + (G_{mm}/G_{uu})}\right] \qquad (4.52)$$

where the dependence on f is omitted to simplify the notation. Thus even if $G_{mm} \ll G_{uu}$, $|H|_a$ will always give a biased estimate when $G_{nn} \neq 0$. However, substitution of Equations 4.42 and 4.43 into Equation 4.50 gives

$$|H|_c = \frac{|G_{uv}|}{G_{uu} + G_{mm}} = |H|\left[\frac{1}{1 + (G_{mm}/G_{uu})}\right] \qquad (4.53)$$

Now $|H_c|$ will give an unbiased estimate of $|H|$ when $G_{mm} \ll G_{uu}$, regardless of the value of G_{nn}.

Equations 4.52 and 4.53 demonstrate conclusively the superiority of the cross-spectrum method over the autospectrum method for estimating system gain factors whenever independent output noise occurs, which will nearly always be the case in practice. These matters are developed further in Chapters 5 and 11.

4.2.2   Input Noise Only

Input noise leads to bias errors in the estimation of $H(f)$ by Equation 4.9 because the denominator will no longer be the true input spectrum. Two situations can arise depending on whether this unknown input noise $m(t)$ passes through the system or is extraneous noise that appears in the measurement of $x(t)$ but does not go through the system. Figure 4.3 illustrates these two situations where the measured values in both cases are $x(t)$ and $y(t)$.

Consider case 1 where input noise passes through the system. Here there could be many physical problems where an unknown and/or unmeasured noise $m(t)$ may be correlated with the measured input $x(t)$. For this case, the finite Fourier transform of Equation 4.1 over record length T yields

$$Y(f, T) = H(f)[X(f, T) + M(f, T)] \tag{4.54}$$

and from Equation 3.46,

$$G_{xy}(f) = H(f)[G_{xx}(f) + G_{xm}(f)] \tag{4.55}$$

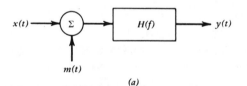

(a)

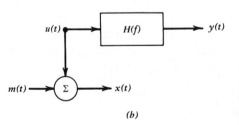

(b)

Figure 4.3   Two input noise situations. (*a*) Noise passes through the system. (*b*) Noise does not pass through the system.

Hence

$$H(f) = \frac{G_{xy}(f)}{G_{xx}(f) + G_{xm}(f)} = \text{true frequency response function} \quad (4.56)$$

$$H_{xy}(f) = \frac{G_{xy}(f)}{G_{xx}(f)} = \text{measured frequency response function} \quad (4.57)$$

It follows from Equations 4.56 and 4.57 that

$$H_{xy}(f) = H(f)\left[1 + \left\{\frac{G_{xm}(f)}{G_{xx}(f)}\right\}\right] \quad (4.58)$$

Thus, in general, $H_{xy}(f)$ will be a biased estimate of $H(f)$ in both gain and phase. However, if $G_{xm}(f) = 0$, which occurs when $x(t)$ and $m(t)$ are uncorrelated, then $H_{xy}(f)$ will provide an unbiased estimate of $H(f)$.

Now consider case 2 where input noise does not pass through the system. Here it is reasonable to assume that $m(t)$ and $u(t)$ are uncorrelated, meaning $m(t)$ and $y(t)$ will be uncorrelated. The following relationships apply:

$$X(f, T) = U(f, T) + M(f, T) \quad (4.59)$$

$$G_{xy}(f) = G_{uy}(f) \quad (4.60)$$

$$G_{xx}(f) = G_{uu}(f) + G_{mm}(f) \quad (4.61)$$

$$H(f) = \frac{G_{uy}(f)}{G_{uu}(f)} = \text{true frequency response function} \quad (4.62)$$

$$H_{xy}(f) = \frac{G_{xy}(f)}{G_{xx}(f)} = \text{measured frequency response function} \quad (4.63)$$

It follows from Equations 4.62 and 4.63 that

$$H_{xy}(f) = H(f)\left[1 + \left\{\frac{G_{mm}(f)}{G_{uu}(f)}\right\}\right]^{-1} \quad (4.64)$$

Thus $H_{xy}(f)$ will always give a biased estimate of $H(f)$ with respect to the gain factor (but not the phase factor), and is a function of the noise-to-signal ratio $G_{mm}(f)/G_{uu}(f)$. However, the cross-spectrum $G_{xy}(f)$ will give an unbiased estimate of $G_{uy}(f)$ in both gain and phase.

4.2.3   Correlated Input and Output Noise

Suppose the noise terms $m(t)$ and $n(t)$ are correlated with each other, but not with the signals, such that their coherence function satisfies

$$0 < \gamma_{mn}^2(f) < 1 \quad (4.65)$$

Define the input noise-to-signal ratio $\alpha(f)$ and the output noise-to-signal ratio $\beta(f)$ by

$$\alpha(f) = \frac{G_{mm}(f)}{G_{uu}(f)} \qquad \beta(f) = \frac{G_{nn}(f)}{G_{vv}(f)} \tag{4.66}$$

Then Equations 4.42 and 4.43 become

$$\begin{aligned}
G_{xx}(f) &= G_{uu}(f)[1 + \alpha(f)] \\
G_{yy}(f) &= G_{vv}(f)[1 + \beta(f)]
\end{aligned} \tag{4.67}$$

$$G_{xy}(f) = G_{uv}(f) + G_{mn}(f) \tag{4.68}$$

The measured coherence function of Equation 4.45 is now given by the general expression

$$\gamma_{xy}^2(f) = \frac{|G_{uv}(f) + G_{mn}(f)|^2}{G_{uu}(f)G_{vv}(f)[1 + \alpha(f)][1 + \beta(f)]} \tag{4.69}$$

Applications of this formula require further assumptions and approximations along the lines of Reference 4.2.

To obtain some definite useful results, assume that the cross-spectrum magnitude $|G_{uv}(f)|$ between the signal terms $u(t)$ and $v(t)$ is such that

$$|G_{mn}(f)| < |G_{uv}(f)| \tag{4.70}$$

Two special cases can then occur corresponding to $|G_{mn}(f)|$ being parallel to $|G_{uv}(f)|$, or $|G_{mn}(f)|$ being perpendicular to $|G_{uv}(f)|$. Bounds for $\gamma_{xy}^2(f)$ in Equation 4.69 will be developed for these two cases.

First, for the case where $|G_{mn}(f)|$ is parallel to $|G_{uv}(f)|$, this situation means

$$|G_{xy}(f)| = |G_{uv}(f) + G_{mn}(f)| = |G_{uv}(f)| \pm |G_{mn}(f)| \tag{4.71}$$

as shown in Figure 4.4. The measured phase angle $\theta_{xy}(f) = \theta_{uv}(f)$. By definition

$$|G_{mn}(f)|^2 = \gamma_{mn}^2(f)G_{mm}(f)G_{nn}(f) \tag{4.72}$$

where $\gamma_{mn}^2(f)$ is the coherence function between $m(t)$ and $n(t)$. Also,

$$|G_{uv}(f)|^2 = G_{uu}(f)G_{vv}(f) \tag{4.73}$$

It follows that

$$\begin{aligned}
|G_{xy}(f)|^2 &= |G_{uv}(f)|^2 \left[1 \pm \frac{|G_{mn}(f)|}{|G_{uv}(f)|}\right]^2 \\
&= G_{uu}(f)G_{vv}(f)[1 \pm \gamma_{mn}(f)[\alpha(f)\beta(f)]^{1/2}]^2 \tag{4.74}
\end{aligned}$$

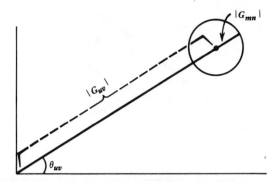

Figure 4.4   Correlated noise with $|G_{mn}(f)|$ parallel to $|G_{uv}(f)|$.

This yields

$$\gamma_{xy}^2(f) = \frac{[1 \pm \gamma_{mn}(f)[\alpha(f)\beta(f)]^{1/2}]^2}{[1 + \alpha(f)][1 + \beta(f)]} \tag{4.75}$$

The maximum reduction in measured coherence will occur by choosing the minus sign in the numerator and by letting $\gamma_{mn}(f)$ approach one. Assuming $\alpha(f) = \beta(f)$, this will give the lowest possible result,

$$\gamma_{xy}^2(f) = \left[\frac{1 - \alpha(f)}{1 + \alpha(f)}\right]^2 \tag{4.76}$$

Values of $\gamma_{xy}^2(f)$ as a function of $\alpha(f)$ are shown in Table 4.1.

Now consider case 2 where $|G_{mn}(f)|$ is perpendicular to $|G_{uv}(f)|$. For this situation

$$|G_{xy}(f)|^2 = |G_{uv}(f)|^2 - |G_{mn}(f)|^2 \tag{4.77}$$

as shown in Figure 4.5. Here the measured phase angle $\theta_{xy}(f) = \theta_{uv}(f) \pm \Delta\theta_{mn}(f)$. By using Equations 4.66 through 4.68, it now follows that

$$|G_{xy}(f)|^2 = G_{uu}(f)G_{vv}(f)[1 - \gamma_{mn}^2(f)\alpha(f)\beta(f)] \tag{4.78}$$

Table 4.1

Minimum $\gamma_{xy}^2(f)$ versus $\alpha(f)$ from Equation 4.76

$\alpha(f)$	0.050	0.100	0.150	0.200	0.250
$\gamma_{xy}^2(f)$	0.819	0.669	0.546	0.444	0.360

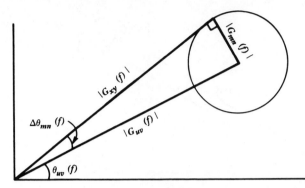

Figure 4.5 Correlated noise with $|G_{mn}(f)|$ perpendicular to $|G_{uv}(f)|$.

Table 4.2

Minimum $\gamma_{xy}^2(f)$ versus $\alpha(f)$ from Equation 4.80

$\alpha(f)$	0.050	0.100	0.150	0.200	0.250
$\gamma_{xy}^2(f)$	0.905	0.818	0.739	0.667	0.600

which yields

$$\gamma_{xy}^2(f) = \frac{1 - \gamma_{mn}^2(f)\alpha(f)\beta(f)}{[1 + \alpha(f)][1 + \beta(f)]} \tag{4.79}$$

As before, the maximum reduction in measured coherence will occur by letting $\gamma_{mn}(f)$ approach one. Assuming $\alpha(f) = \beta(f)$, this will give the lowest possible result,

$$\gamma_{xy}^2(f) = \frac{1 - \alpha(f)}{1 + \alpha(f)} \tag{4.80}$$

which is always larger than Equation 4.76. Values of $\gamma_{xy}^2(f)$ as a function of $\alpha(f)$ are shown in Table 4.2.

4.3 FEEDBACK SYSTEMS

Consider an ideal feedback system with negative feedback as illustrated in Figure 4.6, where $H_1(f)$ is the linear frequency response function of the forward path, $H_2(f)$ is the linear frequency response function of the feedback

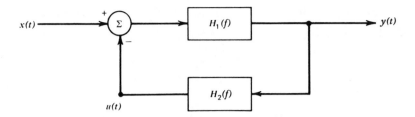

Figure 4.6   Ideal single input/single output feedback system.

path, $x(t)$ is the input, and $y(t)$ is the output. When $H_2(f) = 0$, this model is the usual single input/single output system of Figure 4.1. Some relationships will be listed here to indicate changes that can occur in these situations over previous results which would require new interpretations of the former results. Further useful ideas are in Reference 4.3.

4.3.1   Relationships Without Extraneous Noise

Define $u(t)$ as the output of $H_2(f)$ that would usually not be directly measurable in practice. From Equation 4.1, the finite Fourier transforms of $x(t)$, $y(t)$, and $u(t)$ over a long record length T are related by

$$Y(f, T) = H_1(f)[X(f, T) - U(f, T)]$$
$$U(f, T) = H_2(f)Y(f, T) \tag{4.81}$$

Hence

$$Y(f, T) = H_1(f)[X(f, T) - H_2(f)Y(f, T)] \tag{4.82}$$

which yields

$$H(f) = \frac{Y(f, T)}{X(f, T)} = \frac{H_1(f)}{1 + H_1(f)H_2(f)}$$

The quantity $H(f)$ is the open-loop overall linear frequency response function between $x(t)$ and $y(t)$ that can be computed from measurements only of $x(t)$ and $y(t)$. One would not like to have the product $H_1(f)H_2(f) = -1$ at any frequency. Experimental determination of $H_1(f)$ and $H_2(f)$ is not feasible unless the quantity $u(t)$ can be measured. Of course, if one knows or can assume some form for either $H_1(f)$ or $H_2(f)$, then the other can be determined.

For example, suppose $H_2(f) = c$, a known positive constant. Then

$$H(f) = \frac{H_1(f)}{1 + cH_1(f)} \qquad H_1(f) = \frac{H(f)}{1 - cH(f)} \tag{4.83}$$

For stationary random data, $H(f)$ can be found from the cross-spectra relation

$$H(f) = \frac{G_{xy}(f)}{G_{xx}(f)} \tag{4.84}$$

Hence

$$H_1(f) = \frac{G_{xy}(f)}{G_{xx}(f) - cG_{xy}(f)} \tag{4.85}$$

Clearly, it would be wrong here to interpret $H(f)$ as being the same as $H_1(f)$ unless $c = 0$.

4.3.2   Effects of Extraneous Noise

Suppose unknown extraneous noise $n(t)$ enters into the feedback system as shown in Figure 4.7. In this model, define $v(t)$ as the output of $H_1(f)$ which would no longer be measured. Instead, one measures $y(t) = v(t) + n(t)$ where $n(t)$ is uncorrelated with $v(t)$. Also, $u(t)$, the output of $H_2(f)$, will not be assumed directly measurable. By taking appropriate finite Fourier transforms over a long record length T, the following relationships apply:

$$Y(f, T) = V(f, T) + N(f, T)$$
$$V(f, T) = H_1(f)[X(f, T) - H_2(f)Y(f, T)] \tag{4.86}$$

Hence

$$Y(f, T) = H_1(f)[X(f, T) - H_2(f)Y(f, T)] + N(f, T) \tag{4.87}$$

The solution for the feedback system response is now

$$Y(f, T) = \frac{H_1(f)X(f, T)}{1 + H_1(f)H_2(f)} + \frac{N(f, T)}{1 + H_1(f)H_2(f)} \tag{4.88}$$

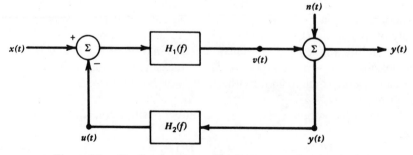

Figure 4.7   Feedback system with extraneous noise at output.

consisting of a signal term plus a noise term. If $H_2(f) = 0$, that is, there is no feedback path, then the response is

$$Y(f, T) = H_1(f)X(f, T) + N(f, T) \tag{4.89}$$

consisting of a different signal term and a different noise term. Both of these relations give the same signal-to-noise ratio at the output, namely,

$$\frac{\text{signal}}{\text{noise}} = \frac{H_1(f)X(f, T)}{N(f, T)} \tag{4.90}$$

which is a function of $H_1(f)$ but not of $H_2(f)$. Thus the output signal-to-noise ratio can be increased by increasing $H_1(f)$ while holding $H_2(f)$ fixed, representing a desirable feature of feedback systems.

Returning now to Equation 4.88, this is of the form

$$Y(f, T) = H(f)X(f, T) + M(f, T) \tag{4.91}$$

where

$$H(f) = \frac{H_1(f)}{1 + H_1(f)H_2(f)} \qquad M(f, T) = \frac{N(f, T)}{1 + H_1(f)H_2(f)} \tag{4.92}$$

For stationary random data, $H(f)$ and $G_{mm}(f)$ can be found from the single input/single output formulas

$$H(f) = \frac{G_{xy}(f)}{G_{xx}(f)} \tag{4.93}$$

$$G_{mm}(f) = [1 - \gamma_{xy}^2(f)]G_{yy}(f) \tag{4.94}$$

The total output spectrum

$$G_{yy}(f) = G_{vv}(f) + G_{mm}(f) \tag{4.95}$$

where

$$G_{vv}(f) = \frac{|H_1(f)|^2 G_{xx}(f)}{|1 + H_1(f)H_2(f)|^2} \tag{4.96}$$

$$G_{mm}(f) = \frac{G_{nn}(f)}{|1 + H_1(f)H_2(f)|^2} \tag{4.97}$$

It follows from Equation 4.31 that

$$\gamma_{xy}^2(f) = \frac{|H_1(f)|^2 G_{xx}(f)}{|H_1(f)|^2 G_{xx}(f) + G_{nn}(f)} \tag{4.98}$$

Thus, $\gamma_{xy}^2(f)$ is independent of $H_2(f)$ in agreement with the fact that the output signal-to-noise ratio is independent of $H_2(f)$.

REFERENCES

4.1 Bendat, J. S., and Piersol, A. G., *Random Data: Analysis and Measurement Procedures*, Wiley-Interscience, New York, 1971.

4.2 Talbot, C. R. S., "Coherence Function Effects on Phase Difference Interpretation," *Journal of Sound and Vibration*, Vol. 39, No. 3, p. 345, 1975.

4.3 Marmarelis, P. Z., and Marmarelis, V. Z., *Analysis of Physiological Systems*, Plenum Press, New York, 1978.

CHAPTER 5

SYSTEM IDENTIFICATION
AND RESPONSE

The most direct applications of the single input/output relationships developed in Chapter 4 are to the estimation of system frequency response functions based on measured single input/output data, and the prediction of system response characteristics from measured input data and known or estimated frequency response functions. System frequency response properties can also be estimated, under certain conditions, from measured response data only. This subject is covered in Chapter 7. The estimation of frequency response functions based on correlated multiple input/output data is detailed in Chapters 8 and 10.

5.1   FREQUENCY RESPONSE FUNCTION CALCULATIONS

Consider a single input/single output system with extraneous noise at the output point only as pictured in Figure 5.1. This corresponds to many practical physical problems where the input measurement $x(t)$ is essentially noise-free while the output measurement $y(t)$ consists of the sum of the ideal linear output $v(t)$ due to $x(t)$, plus all possible deviations $n(t)$. From measurements of $x(t)$ and $y(t)$ only, the system frequency response function $H(f)$ is calculated by using Equation 4.9, namely,

$$H(f) = \frac{G_{xy}(f)}{G_{xx}(f)} \tag{5.1}$$

97

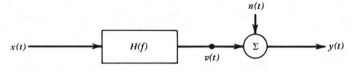

Figure 5.1   Single input/single output system with output noise.

As noted previously in Equations 4.30 and 4.32,

$$G_{vv}(f) = \gamma^2_{xy}(f)G_{yy}(f) \tag{5.2}$$

$$G_{nn}(f) = [1 - \gamma^2_{xy}(f)]G_{yy}(f) \tag{5.3}$$

The quantity $G_{nn}(f)$ is the residual output noise spectrum at $y(t)$ when the linear effects of $x(t)$ are removed from $y(t)$. The quantity $G_{vv}(f)$ is the coherent output spectrum at $y(t)$ due to the linear effects of $x(t)$. The ordinary coherence function can be defined by

$$\gamma^2_{xy}(f) = \frac{G_{vv}(f)}{G_{yy}(f)} \tag{5.4}$$

as previously discussed in Section 4.1.4. Since $G_{vv}(f) \le G_{yy}(f)$ at all f, this provides a simple physical proof that

$$0 \le \gamma^2_{xy}(f) \le 1 \tag{5.5}$$

The coherence function defined by Equation 5.4 is the same as Equation 4.25 when $H(f)$ satisfies Equation 5.1.

5.1.1   *Optimum Character of Calculations*

Without using Equation 5.1, let $H(f)$ be *any* linear frequency response function in Figure 5.1. Then for long records of length T, the equation describing Figure 5.1 is

$$Y(f, T) = H(f)X(f, T) + N(f, T) \tag{5.6}$$

where $Y(f, T)$, $X(f, T)$, and $N(f, T)$ are finite Fourier transforms of $y(t)$, $x(t)$, and $n(t)$, respectively, as defined in Equation 3.44. It follows that

$$N(f, T) = Y(f, T) - H(f)X(f, T) \tag{5.7}$$

and

$$|N(f, T)|^2 = |Y(f, T)|^2 - H(f)X(f, T)Y^*(f, T)$$
$$- H^*(f)X^*(f, T)Y(f, T) + H(f)H^*(f)|X(f, T)|^2 \tag{5.8}$$

From Equations 3.46 and 3.47, taking the expectation of Equation 5.8, multiplying by $(2/T)$, and letting T increase to infinity yields

$$G_{nn}(f) = G_{yy}(f) - H(f)G_{yx}(f) - H^*(f)G_{xy}(f) + H(f)H^*(f)G_{xx}(f) \quad (5.9)$$

This is the form of $G_{nn}(f)$ for *any* $H(f)$. By definition, the *optimum* $H(f)$ will now be defined as that $H(f)$ which minimizes $G_{nn}(f)$ over all possible choices of $H(f)$. This is called the *least-squares estimate*.

The minimization of $G_{nn}(f)$ as a function of $H(f)$ will now be carried out. To simplify the derivation, the dependence on f will be omitted. Thus

$$G_{nn} = G_{yy} - HG_{yx} - H^*G_{xy} + HH^*G_{xx} \quad (5.10)$$

Now let the complex numbers be expressed in terms of their real and imaginary parts as follows:

$$
\begin{aligned}
H &= H_R - jH_I & H^* &= H_R + jH_I \\
G_{xy} &= G_R - jG_I & G_{yx} &= G_R + jG_I
\end{aligned}
\quad (5.11)
$$

Then

$$G_{nn} = G_{yy} - (H_R - jH_I)G_{yx} - (H_R + jH_I)G_{xy} + (H_R^2 + H_I^2)G_{xx} \quad (5.12)$$

To find the form of H that will minimize G_{nn}, one should now set the partial derivatives of G_{nn} with respect to H_R and H_I equal to zero and solve the resulting pair of equations. This gives

$$
\begin{aligned}
\frac{\partial G_{nn}}{\partial H_R} &= -G_{yx} - G_{xy} + 2H_R G_{xx} = 0 \\
\frac{\partial G_{nn}}{\partial H_I} &= jG_{yx} - jG_{xy} + 2H_I G_{xx} = 0
\end{aligned}
\quad (5.13)
$$

which leads to

$$
\begin{aligned}
H_R &= \frac{G_{xy} + G_{yx}}{2G_{xx}} = \frac{G_R}{G_{xx}} \\
H_I &= \frac{j(G_{xy} - G_{yx})}{2G_{xx}} = \frac{G_I}{G_{xx}}
\end{aligned}
\quad (5.14)
$$

Hence the optimum H is the same as Equation 5.1, namely,

$$H = H_R - jH_I = \frac{G_R - jG_I}{G_{xx}} = \frac{G_{xy}}{G_{xx}} \quad (5.15)$$

The optimum H calculated by Equation 5.15, using arbitrary measured records, does not have to be physically realizable; it may only be a theoretically computed result.

It should also be noted here that because of the special form of Equation 5.10, a simple way to derive the same optimum H can be obtained by setting either the partial derivative of G_{nn} with respect to H equal to zero (holding H^* fixed) or setting the partial derivative of G_{nn} with respect to H^* equal to zero (holding H fixed). By this method,

$$\frac{\partial G_{nn}}{\partial H} = -G_{yx} + H^*G_{xx} = 0 \tag{5.16}$$

$$H^* = \frac{G_{yx}}{G_{xx}} \qquad H = \frac{G_{xy}}{G_{xx}} \tag{5.17}$$

Another important property satisfied by the optimum $H(f)$ is revealed by substitution of the optimum system satisfying Equation 5.15 into Equation 5.9. This gives

$$G_{nn}(f) = [1 - \gamma_{xy}^2(f)]G_{yy}(f) \tag{5.18}$$

which leads to

$$G_{vv}(f) = G_{yy}(f) - G_{nn}(f) = \gamma_{xy}^2(f)G_{yy}(f) \tag{5.19}$$

Moreover, using the optimum $H(f)$ shows

$$G_{xv}(f) = H(f)G_{xx}(f) \tag{5.20}$$

and

$$G_{xn}(f) = G_{xy}(f) - H(f)G_{xx}(f) = 0 \tag{5.21}$$

It follows that

$$G_{vn}(f) = H^*(f)G_{xn}(f) = 0 \tag{5.22}$$

Thus $n(t)$ and $v(t)$ will *automatically be uncorrelated* when the optimum $H(f)$ is used to estimate the linear system in Figure 5.1.

5.1.2 Relationships Using External Excitation

In some engineering applications where there is an unmeasurable on-going input $x(t)$, it may be feasible to insert a known random or transient

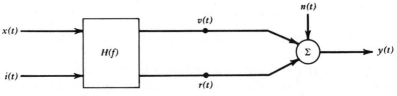

Figure 5.2 System with external excitation and output noise.

external excitation signal $i(t)$ at the desired input point where $i(t)$ is independent of the natural input $x(t)$, as illustrated in Figure 5.2. The quantities in Figure 5.2 have the following meanings:

$x(t) = $ on-going natural input (unmeasured)
$i(t) = $ known external input signal (measured)
$v(t) = $ linear output (unmeasured) caused by $x(t)$
$r(t) = $ linear output (unmeasured) caused by $i(t)$
$n(t) = $ unknown output noise (unmeasured)
$y(t) = v(t) + r(t) + n(t) = $ total output signal (measured)

It is reasonable to assume that $i(t)$ is independent of $x(t)$ and $n(t)$, and $n(t)$ is independent of both $v(t)$ and $r(t)$. The quantity $n(t)$ includes all deviations from the linear model of Figure 5.2 such as nonlinear effects and other uncorrelated inputs.

From Equation 3.44, finite Fourier transforms over long records of length T yield the following results for the quantities in Figure 5.2:

$$R(f, T) = H(f)I(f, T) \qquad V(f, T) = H(f)X(f, T)$$

$$\begin{aligned} Y(f, T) &= R(f, T) + V(f, T) + N(f, T) \\ &= H(f)[I(f, T) + X(f, T)] + N(f, T) \end{aligned} \tag{5.23}$$

By hypothesis, the cross-spectra terms indicated below are zero, that is,

$$G_{ix}(f) = G_{in}(f) = G_{nv}(f) = G_{nr}(f) = 0 \tag{5.24}$$

From the definition in Equations 3.46 and 3.47, it follows that

$$G_{yy}(f) = G_{rr}(f) + G_{vv}(f) + G_{nn}(f) \tag{5.25}$$

$$G_{iy}(f) = H(f)G_{ii}(f) \tag{5.26}$$

If the frequency response function $H(f)$ could be estimated by the usual formula involving the measurement of $x(t)$, namely

$$H(f) = \frac{G_{xy}(f)}{G_{xx}(f)} \tag{5.27}$$

then biased estimates would be obtained when extraneous noise is present in $x(t)$. However, Equation 5.27 is not a feasible procedure when $x(t)$ cannot be measured. In Figure 5.2, by using the known measurable external input signal $i(t)$, Equation 5.26 shows that

$$H(f) = \frac{G_{iy}(f)}{G_{ii}(f)} \tag{5.28}$$

This gives an unbiased estimate of $H(f)$ independent of any measurements or assumptions regarding $x(t)$, and is a recommended procedure in practice. In particular, the external excitation signal $i(t)$ is often taken in the form of bandwidth-limited white noise, where $G_{ii}(f) = K$, a constant, so that $H(f)$ is simply

$$H(f) = \frac{G_{iy}(f)}{K} \tag{5.29}$$

Once the frequency response function $H(f)$ has been determined by Equation 5.28 or 5.29, then $i(t)$ can be removed. Equation 5.23 becomes

$$Y(f, T) = V(f, T) + N(f, T) = H(f)X(f, T) + N(f, T) \tag{5.30}$$

Now the measurement of $y(t)$ alone will give

$$G_{yy}(f) = G_{vv}(f) + G_{nn}(f) = |H(f)|^2 G_{xx}(f) + G_{nn}(f) \tag{5.31}$$

For those cases where $G_{nn}(f) \ll G_{vv}(f)$, one obtains

$$G_{yy}(f) \approx |H(f)|^2 G_{xx}(f) \tag{5.32}$$

It follows that

$$G_{xx}(f) \approx \frac{G_{yy}(f)}{|H(f)|^2} \tag{5.33}$$

which will provide useful information about the autospectrum $G_{xx}(f)$ even though $x(t)$ is not measured. In particular, $x(t)$ could represent the self-noise at the input and $y(t)$ the self-noise at the output due to $x(t)$.

5.1.3   Noise-to-Signal Ratios

With reference to Figure 5.2, define the input and output noise-to-signal ratios $\alpha(f)$ and $\beta(f)$, respectively, by

$$\alpha(f) = \frac{G_{xx}(f)}{G_{ii}(f)} \qquad \beta(f) = \frac{G_{yy}(f) - G_{rr}(f)}{G_{rr}(f)} \tag{5.34}$$

These ratios consider $G_{ii}(f)$ and $G_{rr}(f)$ to be the input and output signals. The numerator $G_{xx}(f)$ in the $\alpha(f)$ term is the input spectrum without $i(t)$, and the numerator $G_{yy}(f) - G_{rr}(f)$ in the $\beta(f)$ term is the output spectrum without $r(t)$. Let

$$x_1(t) = i(t) + x(t) \tag{5.35}$$

Then the autospectra of $x_1(t)$ and $y(t)$ are given by

$$\begin{aligned} G_{x_1 x_1}(f) &= G_{ii}(f) + G_{xx}(f) = G_{ii}(f)[1 + \alpha(f)] \\ G_{yy}(f) &= G_{rr}(f)[1 + \beta(f)] \end{aligned} \tag{5.36}$$

and the cross-spectrum between $x_1(t)$ and $y(t)$ is

$$G_{x_1y}(f) = G_{iy}(f) + G_{xy}(f) = G_{ir}(f) + G_{xv}(f) \qquad (5.37)$$

It follows that

$$
\begin{aligned}
G_{ii}(f) &= \gamma_{ix_1}^2(f)G_{x_1x_1}(f) & G_{xx}(f) &= [1 - \gamma_{ix_1}^2(f)]G_{x_1x_1}(f) \\
G_{rr}(f) &= \gamma_{iy}^2(f)G_{yy}(f) & G_{yy}(f) - G_{rr}(f) &= [1 - \gamma_{iy}^2(f)]G_{yy}(f)
\end{aligned} \qquad (5.38)
$$

The coherence function $\gamma_{iy}^2(f)$ is always measurable, but $\gamma_{ix_1}^2(f)$ may or may not be measurable depending on the knowledge of $x(t)$. From Equations 5.34 and 5.38, one obtains

$$\alpha(f) = \frac{1 - \gamma_{ix_1}^2(f)}{\gamma_{ix_1}^2(f)} \qquad \beta(f) = \frac{1 - \gamma_{iy}^2(f)}{\gamma_{iy}^2(f)} \qquad (5.39)$$

Conversely, this proves that

$$\gamma_{ix_1}^2(f) = \frac{1}{1 + \alpha(f)} \qquad \gamma_{iy}^2(f) = \frac{1}{1 + \beta(f)} \qquad (5.40)$$

Thus $\alpha(f) = \beta(f)$ if and only if $\gamma_{ix_1}^2(f) = \gamma_{iy}^2(f)$. From Equations 5.37 and 5.38, the coherence function $\gamma_{x_1y}^2(f)$ is given by

$$\gamma_{x_1y}^2(f) = \gamma_{ix_1}^2(f)\gamma_{iy}^2(f)\frac{|G_{ir}(f) + G_{xv}(f)|^2}{G_{ii}(f)G_{rr}(f)} \qquad (5.41)$$

When $G_{xv}(f) = 0$, this reduces to the simple result

$$\gamma_{x_1y}^2(f) = \gamma_{ix_1}^2(f)\gamma_{iy}^2(f) \qquad (5.42)$$

5.2   FREQUENCY RESPONSE FUNCTION ESTIMATION ERRORS

Consider the single input/single output system shown in Figure 5.1 where the input $x(t)$ and the output $y(t)$ are stationary (ergodic) time history records measured simultaneously over the finite time interval $0 \le t \le T$. From Equation 5.1, an optimum estimate of the frequency response function of the system is given by

$$\hat{H}(f) = \frac{\hat{G}_{xy}(f)}{\hat{G}_{xx}(f)} \qquad (5.43)$$

where the spectral density functions are estimated from the finite input/output records by the digital procedures detailed in Section 3.4.2. From the developments in Section 4.1.2, the frequency response function can also be estimated for transient records $x(t)$ and $y(t)$ by

$$\hat{H}(f) = \frac{\hat{\mathcal{G}}_{xy}(f)}{\hat{\mathcal{G}}_{xx}(f)} \tag{5.44}$$

where the "energy" spectral density functions are estimated from

$$\hat{\mathcal{G}}_{xy}(f) = T\hat{G}_{xy}(f) \qquad \hat{\mathcal{G}}_{xx}(f) = T\hat{G}_{xx}(f) \tag{5.45}$$

It is assumed in Equations 5.44 and 5.45 that the record length T is sufficiently long to cover all significant values of $x(t)$ and $y(t)$, that is, $x(t) = y(t) = 0$ for $t < 0$ and $t > T$. It is also assumed here that the experiment can be repeated many times using a similar transient excitation to obtain the collection of n_d records needed for the averaging operation in Equations 3.89 and 3.90.

The estimation of frequency response functions from either Equation 5.43 or 5.44 will generally involve both random and bias errors. The key to successful applications is an understanding of these errors and a diligent effort to minimize them. In most cases, the coherence function between the measured input and output will indicate the presence of errors and is helpful in assessing their origin and magnitude. Hence frequency response estimates should *always* be accompanied by coherence function estimates.

5.2.1 Random Errors

Random errors in frequency response function estimates are due to the following sources:

1. Measurement noise in the transducers and instrumentation, and computational noise in the digital calculations.
2. Other unmeasured inputs that contribute to the output and are uncorrelated with the measured input.
3. Nonlinearities in the system between the input and output.

The resulting random error due to these collective sources is directly related to (*a*) the coherence function $\hat{\gamma}_{xy}^2(f)$ calculated between the measured input and output records, and (*b*) the number of averages n_d used in the calculations of the spectral density estimates. Specifically, from Chapter 11, the normalized random error in the estimated gain factor and the standard deviation in the estimated phase factor are given by

$$\varepsilon[|\hat{H}(f)|] \approx \frac{[1 - \hat{\gamma}_{xy}^2(f)]^{1/2}}{|\hat{\gamma}_{xy}(f)|\sqrt{2n_d}} \qquad \sigma[\hat{\phi}(f)] \approx \sin^{-1}\{\varepsilon[|\hat{H}(f)|]\} \tag{5.46}$$

In the laboratory, experiments usually can be designed with a well-defined single input and very low instrumentation noise to obtain coherence functions near unity. In such cases, frequency response estimates can be calculated with a modest random error using a relatively small number of averages. In field experiments, however, it may not be feasible to produce such ideal conditions, particularly when the measured input is a natural excitation rather than a simulated input. Hence the coherence function might be well below unity, meaning a large value of n_d will be required for acceptably accurate results.

The origin of random errors in frequency response estimates from field experiments can often be assessed using the following guidelines from past experience (the reader should be cautioned that these are only rules of thumb with numerous possible exceptions):

1. If $\hat{\gamma}_{xy}^2(f)$ falls broadly over a frequency range where $|\hat{H}(f)|$ is relatively small, this usually suggests extraneous noise at the output $y(t)$ due to measurement noise and/or the contributions of other uncorrelated inputs.
2. If $\hat{\gamma}_{xy}^2(f)$ falls broadly over a frequency range where $|\hat{H}(f)|$ is not near a minimum value and $\hat{G}_{xx}(f)$ is relatively small, then measurement noise at the input $x(t)$ should be suspected. Note that beyond the increased random error, this situation will cause a bias error in the estimates, as discussed in the next section.
3. For sharply peaked gain factor estimates characteristic of lightly damped resonant systems, $\hat{\gamma}_{xy}^2(f)$ will usually peak sharply at the frequencies where $|\hat{H}(f)|$ peaks (frequencies of system resonances) because the signal-to-noise ratio is highest at these frequencies. If $\hat{\gamma}_{xy}^2(f)$ at such frequencies does not sharply peak, or worse yet notches, then system nonlinearities might be suspected. However, resolution bias errors in the spectral estimates are a more likely source of such results, as discussed in the next section.

5.2.2 *Bias Errors*

Bias errors can arise in frequency response function estimates from four primary sources as follows:

1. Extraneous noise in the input measurement that does not pass through the system.
2. Resolution bias errors in the spectral density estimates.
3. Nonlinear system parameters.
4. Other unmeasured inputs that contribute to the output and are correlated with the measured input.

The first error due to input measurement noise is detailed in Section 4.2.2. Specifically, assuming the measured input $x(t)$ includes measurement noise $m(t)$ plus the true system input $u(t)$, the expected value of the estimated frequency response function is given by Equation 4.64 as

$$E[\hat{H}(f)] = H(f)\left\{1 + \left[\frac{G_{mm}(f)}{G_{uu}(f)}\right]\right\}^{-1} \qquad (5.47)$$

that is, input measurement noise will always cause the true frequency response function to be underestimated. Hence it is very important that the input measurement $x(t)$ be as noise free as feasible. If the input noise is an actual excitation that passes through the system and is uncorrelated with the input $u(t)$ of interest, then no bias occurs as demonstrated in Section 4.2.2.

The second problem due to resolution bias errors in the spectral density estimates $\hat{G}_{xx}(f)$ and $\hat{G}_{xy}(f)$ will be most prevalent at peaks and notches in magnitude of the estimated frequency response function. These spectral density resolution errors are developed in Chapter 11 and summarized in Section 3.4.3. Such errors can be suppressed by obtaining properly resolved estimates of the spectral density functions, that is, by making the resolution bandwidth B_e sufficiently narrow to define the spectral peaks and notches accurately.

The third indicated source of bias results from a violation of the assumption that the system is linear. The application of Equation 5.43 or 5.44 to nonlinear systems will produce only a linear approximation of the system response characteristics. However, as proved in Section 5.1.1, the result will be the best possible linear approximation in the least-squares sense for the frequency response function under the specified input conditions. This fact constitutes a strong argument for estimating frequency response functions of systems with suspected nonlinear properties using natural inputs or accurate laboratory simulations of natural inputs, rather than arbitrary laboratory inputs.

The final error due to other correlated inputs involves matters to be developed in Chapter 8. To illustrate the form of the error, consider a two-input situation where $x(t)$ is the measured input and $z(t)$ is a second unmeasured input that together produce the output $y(t)$. The expected value of the frequency response function $\hat{H}(f)$ estimated by Equation 5.43 compared to the true frequency response function $H(f)$ in this case will be shown in Chapter 8 to be

$$\frac{E[\hat{H}(f)]}{H(f)} = \frac{1 - \gamma_{xz}^2(f)}{1 - [G_{xz}(f)G_{zy}(f)/G_{zz}(f)G_{xy}(f)]} \qquad (5.48)$$

Note that if the measured and unmeasured inputs $x(t)$ and $z(t)$ are uncorrelated, then $\gamma_{xz}^2(f) = 0$ and the right side of Equation 5.48 reduces to unity.

This demonstrates that unmeasured inputs which are uncorrelated with the measured input cause no bias error since they appear at the output as extraneous noise. They do, of course, contribute to the random error in frequency response function estimates, as discussed in the preceding section.

It should be mentioned that there is a minor statistical bias error inherent in the estimation procedure of Equations 5.43 and 5.44 due to the fact that, in general,

$$E[\hat{H}(f)] = E\left[\frac{\hat{G}_{xy}(f)}{\hat{G}_{xx}(f)}\right] \neq \frac{E[\hat{G}_{xy}(f)]}{E[\hat{G}_{xx}(f)]} \tag{5.49}$$

Hence $E[\hat{H}(f)] \neq H(f)$. However, this is not a problem in practice because the number of averages n_d required to suppress random errors in Equation 5.46 will generally make the inherent bias suggested by Equation 5.49 negligible.

The existence of bias errors in frequency response function estimates due to input noise, inadequate spectral resolution, and nonlinear effects will all produce indicative anomalies in the measured coherence function $\hat{\gamma}_{xy}^2(f)$. Specifically, the following general guidelines apply:

1. If $\hat{\gamma}_{xy}^2(f)$ falls broadly over a frequency range where $|\hat{H}(f)|$ is not near a minimum value and $\hat{G}_{xx}(f)$ is relatively small, then measurement noise at the input $x(t)$ should be suspected.
2. If $\hat{\gamma}_{xy}^2(f)$ notches sharply at a frequency where $|\hat{H}(f)|$ displays a sharp peak or notch, then inadequate spectral resolution in the analysis is the most likely cause, although nonlinearities can produce similar results at peaks in $|\hat{H}(f)|$.
3. To distinguish between resolution problems and nonlinearities, the measurements should be repeated with an improved spectral resolution. An increased value of $\hat{\gamma}_{xy}^2(f)$ will confirm a resolution problem. Otherwise, nonlinearities should be studied.

Bias errors due to other unmeasured inputs that are correlated with the measured input $x(t)$ will not necessarily be reflected in $\hat{\gamma}_{xy}^2(f)$. This problem must be addressed in the context of multiple input/output systems, discussed in Chapter 8.

5.2.3 Illustrations of Errors

To illustrate some of the problems that can arise in frequency response function estimates using single input/output data, consider an early laboratory experiment performed by Barnoski [5.2], as outlined in Figure 5.3. A thin cantilever beam was excited at the fixed end through a heavy mounting

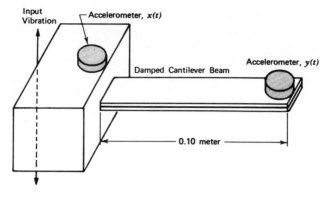

Figure 5.3   Cantilever beam vibration experiment.

fixture by broad-band random vibration covering a frequency range from near zero to 1200 Hz. An accelerometer on the mounting fixture provided the input $x(t)$ while a second accelerometer on the end of the beam produced the output $y(t)$. The data were tape-recorded for analysis which limited the measurement signal-to-noise ratio to about 45 dB. The analysis was performed with a resolution bandwidth of $B_e \approx 5$ Hz and $n_d = 29$ averages.

The computed input and output autospectra estimates $\hat{G}_{xx}(f)$ and $\hat{G}_{yy}(f)$ as well as the cross-spectrum magnitude $|\hat{G}_{xy}(f)|$ are shown in Figure 5.4. Three features of the input spectrum $\hat{G}_{xx}(f)$ in Figure 5.4a should be noted. First, the spectral values fall off rapidly at frequencies below 25 Hz and above 700 Hz due to limitations of the excitation source. Second, there is a periodic component present at 60 Hz due to power-line pickup by the instrumentation. Third, there are substantial uncertainty fluctuations (random error) apparent in the spectral estimate due to the relatively small number of averages; from Table 3.1, $n_d = 29$ means $\varepsilon_r = 0.19$ or about 20%. The output spectrum $\hat{G}_{yy}(f)$ in Figure 5.4b and the cross-spectrum magnitude $|\hat{G}_{xy}(f)|$ in Figure 5.4c display greater spectral variations, but otherwise they are similar to the input spectrum in that the spectral values fall off at low and high frequencies, the 60 Hz pickup is present, and random errors appropriate for $n_d = 29$ averages are apparent.

The computed coherence function $\hat{\gamma}_{xy}^2(f)$ is shown in Figure 5.5. Note that $\hat{\gamma}_{xy}^2(f)$ is near unity at most frequencies, as it should be for an ideal input/output problem. Furthermore, the uncertainty fluctuations in $\hat{\gamma}_{xy}^2(f)$ diminish as $\hat{\gamma}_{xy}^2(f)$ approaches unity, exactly as predicted by the appropriate error formula in Table 3.1. Also note that $\hat{\gamma}_{xy}^2(f)$ is near unity at $f = 60$ Hz, in spite of the obvious power-line contamination of the input/output data seen in Figure 5.4. This tells us that the power-line pickup causes no error in the

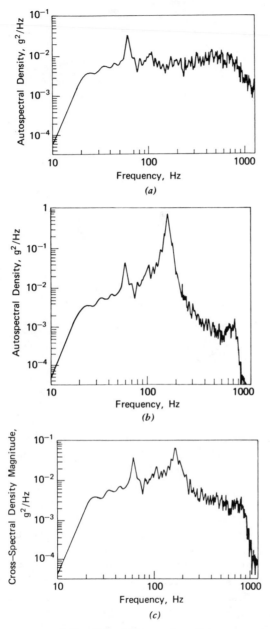

Frequency resolution $B_e = 5$ Hz, number of averages $n_d = 29$

Figure 5.4 Input and output spectra for cantilever beam experiment. (*a*) Input autospectrum. (*b*) Output autospectrum. (*c*) Cross-spectrum magnitude. (The data in Figures 5.4 through 5.6 are taken from Reference 2.1 with the permission of The American Institute of Aeronautics and Astronautics and the author.)

109

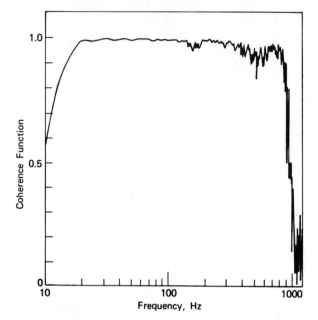

Figure 5.5   Coherence function for cantilever beam experiment.

input/output calculations, probably because it was in the excitation signal and passed through the system as discussed in Section 4.2.2. On the other hand, $\hat{\gamma}_{xy}^2(f)$ falls significantly below unity, warning of an error in the resulting frequency response function estimates in four frequency regions: (a) $f < 20$ Hz, (b) 150 Hz $< f < 190$ Hz, (c) 400 Hz $< f < 700$ Hz, and (d) $f > 800$ Hz.

The estimated frequency response function gain $|\hat{H}(f)|$ and phase $\hat{\phi}(f)$ are shown in Figure 5.6. Note that $|\hat{H}(f)|$ has distinct peaks and $\hat{\phi}(f)$ displays rapid 180-degree phase shifts at two frequencies, namely, $f = 165$ Hz and $f = 820$ Hz. From Section 1.3.3, these results are indicative of the first two normal modes of the cantilever beam. Now the coherence data in Figure 5.5 advise us that the estimate $\hat{H}(f)$ involves either random or bias errors in four separate frequency regions. First, for $f < 20$ Hz, $|\hat{H}(f)|$ is not near a minimum value, but $\hat{G}_{xx}(f)$ in Figure 5.4a is relatively small. From the discussions in Section 5.2.2, this suggests a possible bias error in $|\hat{H}(f)|$ due to measurement noise in $x(t)$. The results in Figure 5.6a confirm this is the case since $|\hat{H}(f)|$ falls below a gain of unity at $f < 20$ Hz even though basic physics requires that $|\hat{H}(f)| \to 1$ as $f \to 0$. Next, the small notch in $\hat{\gamma}_{xy}^2(f)$ around $f = 165$ Hz coincides with a sharp peak in $|\hat{H}(f)|$. Hence it probably represents a resolution bias error in the spectral estimates. The third drop in coherence where 400 Hz $< f < 700$ Hz coincides with a minimum in

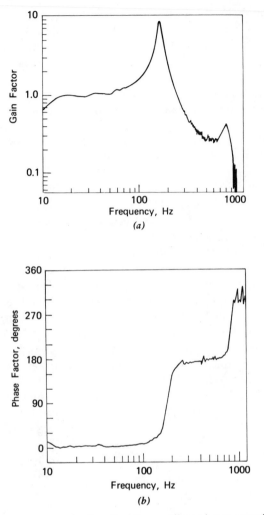

Figure 5.6 Frequency response function for cantilever beam experiment. (*a*) Gain factor. (*b*) Phase factor.

$|\hat{H}(f)|$. From the discussions in Section 5.2.1, this suggests a modest random error in accordance with Equation 5.46, but no bias error, due to small amounts of measurement noise at the output. From Figure 5.6, small uncertainty fluctuations are clearly apparent in both $|\hat{H}(f)|$ and $\hat{\phi}(f)$ over this frequency range, as predicted by Equation 5.46. Finally, the dramatic drop in $\hat{\gamma}^2_{xy}(f)$ at $f > 800$ Hz coincides with a rapid fall in $|\hat{H}(f)|$ above this frequency and obviously reflects massive contamination of the output signal by measurement noise. From Figure 5.6, the random errors in both $|\hat{H}(f)|$ and $\hat{\phi}(f)$ increase above 800 Hz to a point where the estimates become meaningless as $\hat{\gamma}_{xy}^2(f)$ approaches zero, exactly as predicted by Equation 5.46.

To further illustrate the specific problem of resolution bias errors, consider a second laboratory experiment involving a panel structure, as outlined in Figure 5.7. Broad-band random excitation was applied to the center of the panel through a transducer that measured the input force and the output acceleration. The data were analyzed on line with a measurement signal-to-noise ratio in excess of 60 dB. The analysis was performed using a resolution bandwidth of $B_e = 2$ Hz and $n_d = 256$ averages.

The estimated gain factor $|\hat{H}(f)|$ of the frequency response function between $x(t) = $ force and $y(t) = $ acceleration, called inertance, and the associated coherence function estimate $\hat{\gamma}_{xy}^2(f)$ calculated over a frequency range from 20 to 600 Hz are shown in Figure 5.8. Note that $|\hat{H}(f)|$ displays sharp peaks and notches representing several normal modes of the panel. Further note that $\hat{\gamma}_{xy}^2(f)$ is near unity at most frequencies. With the strong coherence values and $n_d = 256$ averages, there is no significant random error in the estimates as predicted by Equation 5.46. There are, however, notches in $\hat{\gamma}_{xy}^2(f)$ at frequencies that coincide with peaks and notches in $|\hat{H}(f)|$. These results are typical of marginal resolution in the spectral density estimates.

To confirm that the notching of $\hat{\gamma}_{xy}^2(f)$ is due to resolution bias errors, the analysis was repeated over a frequency range covering the first peak and notch in $|\hat{H}(f)|$, specifically, from 20 to 150 Hz with a resolution of $B_e = 0.3$ Hz rather than 2 Hz as before. The results are shown in Figure 5.9. Note that $\hat{\gamma}_{xy}^2(f)$ is now near unity at all frequencies including those where notches occurred in the $B_e = 2$ Hz resolution analysis in Figure 5.8, fully confirming that the notches were due to resolution bias errors. It should be observed,

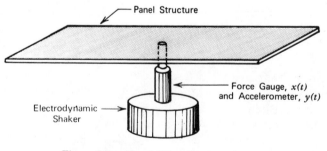

Figure 5.7　Panel Vibration experiment.

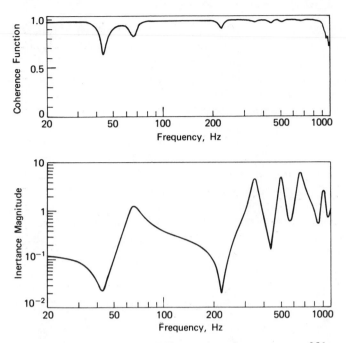

Frequency resolution $B_e = 2$ Hz, number of averages $n_d = 256$

Figure 5.8 Coherence function and gain factor for panel experiment.

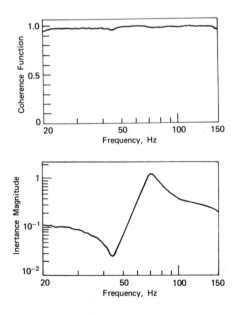

Frequency resolution $B_e = 0.3$ Hz, Number of averages $n_d = 256$

Figure 5.9 Coherence function and gain factor for panel experiment with improved spectral resolution.

113

however, that the improved resolution does not significantly alter the value of $|\hat{H}(f)|$. This demonstrates the extreme sensitivity of the coherence function to relatively minor resolution bias errors.

The spectral estimates used to produce the results in Figures 5.8 and 5.9 were computed with a conventional "Hanning" tapering procedure for side-lobe suppression, as discussed in Section 3.4.2. To illustrate the impact that tapering can have on frequency response function estimates, $|\hat{H}(f)|$ and $\hat{\gamma}^2_{xy}(f)$ were again computed with $B_e = 2$ Hz, but with a rectangular time window in the spectral calculations (no tapering). The results are shown in comparison to the same calculations with tapering in Figure 5.10. Note that the notching of $\hat{\gamma}^2_{xy}(f)$ at peaks and valleys in $|\hat{H}(f)|$ has dramatically increased. Further note that $|\hat{H}(f)|$ is now noticeably different, indicating that the tapering operation in the spectral calculations can make a significant contribution to the accuracy of the estimated gain factor in those cases where the basic spectral resolution is marginal.

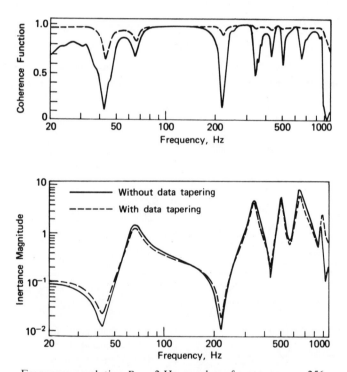

Frequency resolution $B_e = 2$ Hz, number of averages $n_d = 256$

Figure 5.10 Coherence function and gain factor for panel experiment with and without data tapering.

5.3   SYSTEM RESPONSE PREDICTIONS

Again referring to the single input/output system shown in Figure 5.1, consider the situation where the frequency response function $H(f)$ is known or has been estimated by the procedures discussed in Section 5.2. Furthermore, assume that the input $x(t)$ is known or has been measured in terms of its autospectrum $G_{xx}(f)$. From Equations 4.3 and 4.8, the autocorrelation and autospectrum of the response $y(t)$ are given by

$$R_{yy}(\tau) = \int_0^\infty \int_0^\infty h(\alpha)h(\beta)R_{xx}(\tau + \alpha - \beta)\, d\alpha\, d\beta \tag{5.50}$$

$$G_{yy}(f) = |H(f)|^2 G_{xx}(f) \tag{5.51}$$

These results can readily be generalized for distributed inputs using a classical normal mode description of the system, as will be detailed shortly.

5.3.1   *Single Degree-of-Freedom System*

Consider first the simple damped spring-mass (single degree-of-freedom) system shown in Figure 1.8, where the input $x(t) = F(t)$ is force and the output $y(t)$ is displacement. The frequency response function of this system is given by Equation 1.54 as

$$H(f) = \frac{1/k}{1 - (f/f_n)^2 + j2\zeta f/f_n} \tag{5.52}$$

where

$$f_n = \frac{1}{2\pi}\sqrt{\frac{k}{m}} = \text{undamped natural frequency}$$

$$\zeta = \frac{c}{2\sqrt{km}} = \text{damping ratio} \tag{5.53}$$

and m, c, and k are the mass, damping coefficient, and spring rate, respectively, of the system. The unit impulse response function of the system is given by the inverse Fourier transform of Equation 5.52, which for $\zeta < 1$ is

$$h(\tau) = Ae^{-2\pi f_n \zeta \tau} \sin 2\pi f_d \tau \qquad \tau > 0 \tag{5.54}$$

where

$$f_d = f_n\sqrt{1 - \zeta^2} = \text{damped natural frequency}$$

$$A = \frac{2\pi f_n^2}{kf_d} = \frac{2\pi f_n}{k\sqrt{1 - \zeta^2}}$$

Hence the autocorrelation and autospectral density functions of the system response $y(t)$ are given from Equations 5.50 and 5.51 as

$$R_{yy}(\tau) = A^2 \int_0^\infty \int_0^\infty e^{-2\pi f_n \zeta(\alpha+\beta)} \sin 2\pi f_d \alpha \sin 2\pi f_d \beta \, R_{xx}(\tau + \alpha - \beta) \, d\alpha \, d\beta$$

(5.55)

$$G_{yy}(f) = \frac{G_{xx}(f)/k^2}{[1 - (f/f_n)^2]^2 + [2\zeta f/f_n]^2}$$

(5.56)

To illustrate a specific case, assume the input $x(t)$ is representative of a stationary random process $\{x(t)\}$ with a spectral density function $G_{xx}(f) = G$, a constant. For this case, Equations 5.55 and 5.56 become

$$R_{yy}(\tau) = \frac{G\pi f_n e^{-2\pi f_n \zeta |\tau|}}{4\zeta k^2} \left(\cos 2\pi f_d \tau + \frac{\zeta}{\sqrt{1-\zeta^2}} \sin 2\pi f_d |\tau| \right)$$

$$\approx \frac{G\pi f_n e^{-2\pi f_n \zeta |\tau|}}{4\zeta k^2} \cos 2\pi f_n \tau \qquad \text{for } \zeta \ll 1$$

(5.57)

$$G_{yy}(f) = \frac{G/k^2}{[1 - (f/f_n)^2]^2 + [2\zeta f/f_n]^2}$$

(5.58)

These input and output functions along with the system properties are illustrated in Figure 5.11. The total mean square value of the response is given by Equation 3.33 as

$$\psi_y^2 = R_{yy}(0) = \int_0^\infty G_{yy}(f) \, df = \frac{G\pi f_n}{4\zeta k^2}$$

(5.59)

For the common situation where $\zeta \ll 1$, Equations 5.57 through 5.59 will provide reasonably accurate approximations of the system response as long as $G_{xx}(f) \approx G$ over the frequency range $(1 - 6\zeta)f_n < f < (1 + 6\zeta)f_n$.

5.3.2 Distributed Systems

The response prediction procedures in the previous section are easily extended to arbitrary distributed systems subjected to distributed inputs if the system can be described in terms of its normal modes, as discussed for mechanical systems in Section 1.3.4. Specifically, consider a one-dimensional structure subjected to a distributed excitation $p(x, t)$, as shown in Figure 5.12. From Equation 1.60, the response of the structure can be described at any point ξ along its length by

$$y(\xi, t) = \sum_i \phi_i(\xi) q_i(t) \qquad i = 1, 2, 3, \ldots$$

(5.60)

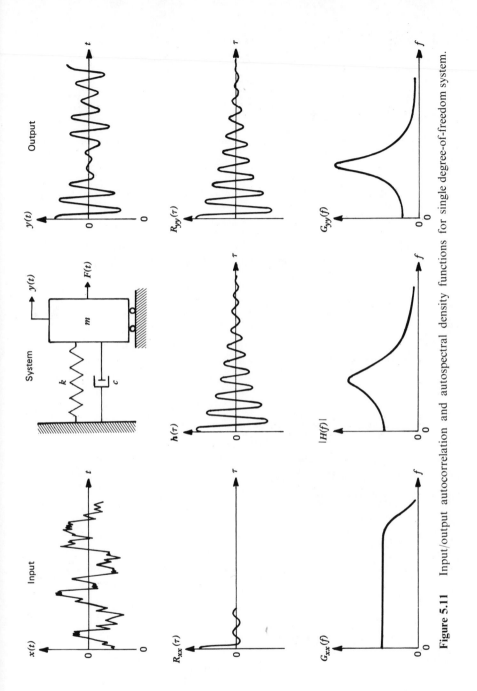

Figure 5.11 Input/output autocorrelation and autospectral density functions for single degree-of-freedom system.

117

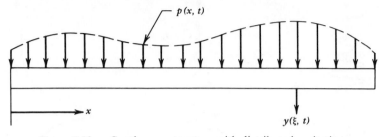

Figure 5.12   Continuous structure with distributed excitation.

where $\phi_i(\xi)$ is the mode shape of the ith normal mode and $q_i(t)$ is the generalized coordinate of the ith normal mode. The function $q_i(t)$ satisfies the differential equation

$$M_i \frac{d^2 q_i(t)}{dt^2} + C_i \frac{dq_i(t)}{dt} + K_i q_i(t) = \int_0^l \phi_i(x)p(x, t)\, dx \qquad (5.61)$$

where M_i, C_i, and K_i are the generalized mass, damping, and stiffness, respectively, of the structure, as defined in Equation 1.61. The autospectrum of the structural response $y(\xi, t)$ is obtained as follows: The finite Fourier transform of Equation 5.60 over a long record length T yields

$$Y(\xi, f, T) = \int_0^T y(\xi, t)e^{-j2\pi ft}\, dt = \sum_i \phi_i(\xi)Q_i(f, T) \qquad (5.62)$$

where $Q_i(f, T)$ is the finite Fourier transform of $q_i(t)$. A similar finite Fourier transform of Equation 5.61 gives

$$-(2\pi f)^2 M_i Q_i(f, T) + j(2\pi f)C_i Q_i(f, T) + K_i Q_i(f, T)$$

$$= \int_0^l \phi_i(x)P(x, f, T)\, dx \qquad (5.63)$$

where $P(x, f, T)$ is the finite Fourier transform of $p(x, t)$. It follows that

$$Q_i(f, T) = H_i(f) \int_0^l \phi_i(x)P(x, f, T)\, dx \qquad (5.64)$$

where substitutions from Equation 5.52 yield

$$H_i(f) = \frac{1/K_i}{[1 - (f/f_i)^2 + j2\zeta_i f/f_i]} \qquad (5.65)$$

Note that the frequency response function $H_i(f)$ for each normal mode, $i = 1, 2, 3, \ldots$, is of the same form as $H(f)$ for the single degree-of-freedom system defined in Equation 5.52. Now, substituting Equation 5.64 into 5.62 yields

$$Y(\xi, f, T) = \sum_i \phi_i(\xi) H_i(f) \int_0^l \phi_i(x) P(x, f, T) \, dx \qquad (5.66)$$

From Equation 3.47, the autospectrum of the output $y(\xi, t)$ is given by

$$G_{yy}(\xi, f) = \lim_{T \to \infty} \frac{2}{T} E[Y^*(\xi, f, T) Y(\xi, f, T)] \qquad (5.67)$$

Hence by substitution from Equation 5.66,

$$G_{yy}(\xi, f) = \sum_i \sum_k \phi_i(\xi)\phi_k(\xi) H_i^*(f) H_k(f) \int_0^l \int_0^l \phi_i(\alpha)\phi_k(\beta) G_{p_\alpha p_\beta}(\alpha, \beta, f) \, d\alpha \, d\beta$$

$$(5.68)$$

where

$$G_{p_\alpha p_\beta}(\alpha, \beta, f) = \lim_{T \to \infty} \frac{2}{T} E[P^*(\alpha, f, T) P(\beta, f, T)] \qquad (5.69)$$

is the cross-spectral density function of the load $p(x, t)$ between all points α and β in space.

The double integral expression in Equation 5.68, with proper normalization, is called the *cross-joint acceptance function* of the structure. It is usually written as

$$j_{ik}^2(f) = \frac{1}{l^2 G_r(f)} \int_0^l \int_0^l \phi_i(\alpha)\phi_k(\beta) G_{p_\alpha p_\beta}(\alpha, \beta, f) \, d\alpha \, d\beta \qquad (5.70)$$

where l is the length of the structure and $G_r(f)$ is a reference autospectrum of the input load density. Equation 5.68 then becomes

$$G_{yy}(\xi, f) = l^2 G_r(f) \sum_i \sum_k \phi_i(\xi)\phi_k(\xi) H_i^*(f) H_k(f) j_{ik}^2(f) \qquad (5.71)$$

The term $j_{ik}^2(f)$ describes how the input couples to the structure over its length. It can sometimes (but not always) be assumed that the cross terms in Equation 5.71 are negligible, in which case

$$G_{yy}(\xi, f) = l^2 G_r(f) \sum_i \phi_i^2(\xi) |H_i(f)|^2 j_i^2(f) \qquad (5.72)$$

where

$$j_i^2(f) = \frac{1}{l^2 G_r(f)} \int_0^l \int_0^l \phi_i(\alpha)\phi_i(\beta) G_{p_\alpha p_\beta}(\alpha, \beta, f) \, d\alpha \, d\beta \qquad (5.73)$$

Furthermore, it can often be assumed that the input $p(x, t)$ is homogeneous (stationary in space) so that

$$j_i^2(f) = \frac{1}{l^2 G_r(f)} \int_0^l \int_0^l \phi_i(\alpha)\phi_i(\beta)G_{p_{\Delta\xi}}(\Delta\xi, f)\, d\alpha\, d\beta \qquad (5.74)$$

where $\Delta\xi = \alpha - \beta$. The above results readily extend to two or more dimensions as required [5.3].

In practice, a number of additional simplifications to Equation 5.68 are often made. For those cases where the system response is dominated by the higher order modes, which are heavily overlapped and cannot be accurately defined by either analytical or experimental procedures, prediction techniques based on statistical averages over space and narrow frequency bands (called Statistical Energy Analysis or SEA procedures) are sometimes used [5.4]. From an empirical viewpoint, since measurements are made at points in space rather than over continuous areas, it is common to reduce distributed input/output systems to multiple point input/output models for analysis and prediction purposes. These matters are covered in Chapters 8 and 10.

REFERENCES

5.1 Bendat, J. S., and Piersol, A. G., *Random Data: Analysis and Measurement Procedures*, Wiley-Interscience, New York, 1971.

5.2 Barnoski, R. L., "Ordinary Coherence Functions and Mechanical Systems," *AIAA Journal of Aircraft*, Vol. 6, No. 4, p. 372, August 1969.

5.3 Wilby, J. F., "The Response of Simple Panels to Turbulent Boundary Layer Excitation," AFFDL-TR-67-70, Wright-Patterson Air Force Base, Ohio, 1967.

5.4 Lyon, R. H., *Statistical Energy Analysis of Dynamical Systems*, The MIT Press, Cambridge, Massachusetts, 1975.

CHAPTER 6

PROPAGATION PATH IDENTIFICATION

A class of problems of common interest, particularly in acoustic noise and vibration control engineering, involves a situation where energy propagates from one location to another through a collection of r number of parallel paths, as illustrated in Figure 6.1. A frequency response function calculation between the input and output measurements in this case will produce a correct overall linear relationship, but it will not in itself identify the contributions of individual paths. To resolve such problems, it is first necessary to distinguish clearly between frequency dispersive and non-dispersive propagations, that is, whether the propagation speed is or is not a function of frequency. Some types of energy propagation are dispersive, for example, surface waves on the ocean and bending waves in structures. In many other cases, however, the propagation can be assumed nondispersive for analysis purposes, for example, electromagnetic radiation and longitudinal (compressive) waves in structures including the air and water (acoustic noise).

6.1 NONDISPERSIVE PROPAGATION, INPUT/OUTPUT DATA

Consider the situation where the propagation paths in Figure 6.1 are nondispersive, that is, the propagation velocity through each path is independent of frequency. Assume that each path has a uniform gain factor H_k,

121

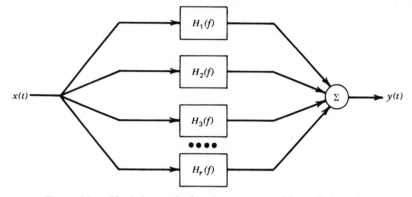

Figure 6.1   Single input/single output system with multiple paths.

$k = 1, 2, 3, \ldots, r$. It follows that an input $x(t)$ from a stationary (ergodic) random process $\{x(t)\}$ will produce an output

$$y(t) = H_1 x(t - \tau_1) + H_2 x(t - \tau_2) + \cdots + H_r x(t - \tau_r)$$

$$= \sum_{k=1}^{r} H_k x(t - \tau_k)$$

(6.1)

6.1.1   Cross-Correlation Measurements

From Equations 3.16 and 3.18, the cross-correlation function between the input $x(t)$ and the output $y(t)$ in Equation 6.1 is given by

$$R_{xy}(\tau) = \lim_{T \to \infty} \frac{1}{T} \int_0^T x(t)[H_1 x(t - \tau_1 + \tau) + H_2 x(t - \tau_2 + \tau)$$

$$+ \cdots + H_r x(t - \tau_r + \tau)] \, dt$$

$$= \sum_{k=1}^{r} H_k R_{xx}(\tau - \tau_k)$$

(6.2)

In other words, the cross-correlation between $x(t)$ and $y(t)$ will appear as a sum of autocorrelation functions of $x(t)$, each offset by τ_k along the time delay scale, and each with a magnitude H_k, $k = 1, 2, 3, \ldots, r$. Hence if the time delay differences among the paths are sufficient to permit a clean separation of the individual peaks, each peak can be used to identify the propagation time and magnitude of the energy through each path. In most cases, the propagation velocity c_k, $k = 1, 2, 3, \ldots, r$, is known, so a length for each path can be calculated from

$$d_k = c_k \tau_k \qquad k = 1, 2, 3, \ldots, r \qquad (6.3)$$

From the geometry of the situation, the distances given by Equation 6.3 will usually identify the specific physical path of propagation.

To illustrate these principles, consider the three path acoustic propagation problem outlined in Figure 6.2a. Acoustic noise covering a frequency range

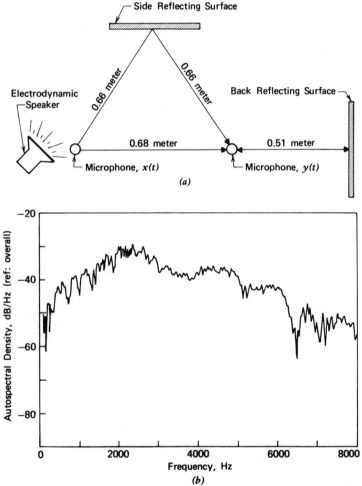

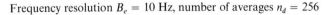

Frequency resolution $B_e = 10$ Hz, number of averages $n_d = 256$

Figure 6.2 Multiple path acoustic experiment. (*a*) Experimental setup. (*b*) Measured autospectrum of output for direct path only with $B = 8000$ Hz.

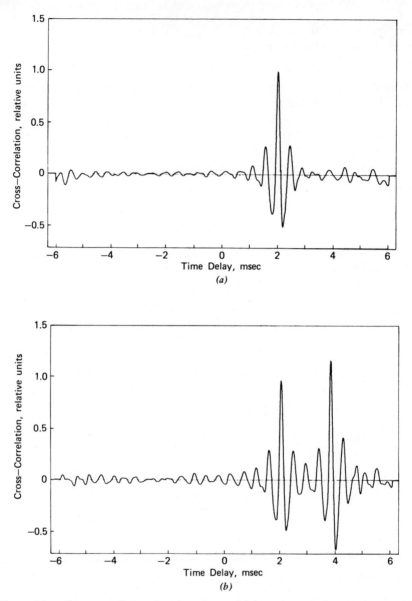

Figure 6.3 Cross-correlation functions for multiple path acoustic experiment with $B = 8000$ Hz. (a) Direct path only. (b) Side reflection present.

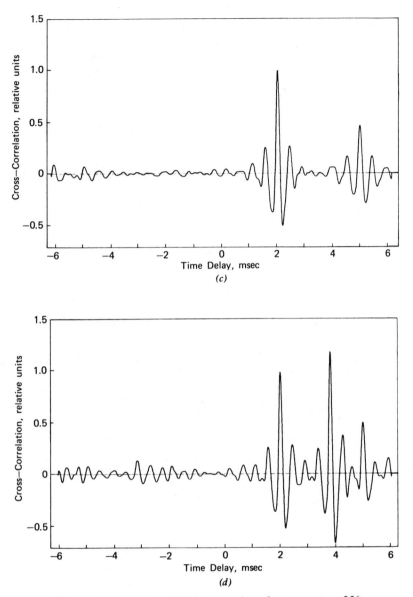

Time resolution $\tau_e = 0.012$ msec, number of averages $n_d = 256$

Figure 6.3 Cross-correlation functions for multiple path acoustic experiment with $B = 8000$ Hz. (*c*) Back reflection present. (*d*) Side and back reflections present.

from near zero to 8000 Hz was applied from a speaker. The autospectrum of the output microphone signal with neither the side nor back reflecting surface present is shown in Figure 6.2b. The cross-correlations computed between the input and output microphones are presented in Figure 6.3 for various reflection configurations. Figure 6.3a shows the cross-correlation function with neither reflecting surface present. Note that the correlation peak occurs at $\tau_1 = 2.0$ msec, which corresponds correctly to the time required for sound to propagate over a distance of $d_1 = 0.68$ meter with a speed in air of about $c = 340$ meters/sec. Furthermore, note that the shape of the correlation function is consistent with the theoretical form computed for band-limited white noise in Equation 3.63.

Figure 6.3b shows the cross-correlation function with only the side reflecting surface present. The correlation peak for the direct path at $\tau_1 = 2.0$ msec remains unchanged and a second peak of similar form is added at $\tau_2 = 3.9$ msec, which is correct for the side reflection propagation distance of $d_2 = 1.32$ meters. The contribution of the reflected noise is greater than the direct path noise because the speaker, which is somewhat directional, was actually pointed at the side reflecting surface. Figure 6.3c shows the cross-correlation with the back reflecting surface only. Again, the direct path correlation peak is unaltered and a second peak of similar form is added at $\tau_3 = 5.0$ msec, which corresponds to the total path length for the back reflection of $d_3 = 1.70$ meters. Finally, the cross-correlation function with both the side and back reflections present is shown in Figure 6.3d. The three correlation peaks corresponding to the three individual paths are clearly apparent at the appropriate time delays. All of these calculations were performed using a time resolution of $\tau_e = \Delta t = 0.012$ msec and $n_d = 256$ averages.

From the discussions in Section 3.3.1, it is clear that the ability to identify correlation peaks corresponding to individual paths will become increasingly difficult as the bandwidth of the data is reduced for a given time delay difference. To demonstrate this point, the experiment outlined in Figure 6.2 was repeated with a nondirectional source having a bandwidth of about 800 Hz on a center frequency of 1000 Hz. The resulting cross-correlation function between the input and output microphones with the side reflection only is shown in Figure 6.4. Note that the two paths, which were clearly apparent for the 8000 Hz bandwidth data in Figure 6.3b, are now only marginally discernible. In the limiting case of a sinusoidal source where the correlation for each path is a cosine function as given by Equation 3.61, no distinction among different paths is possible. As a rule of thumb, if the output data have a bandwidth of B, the propagation time delays for any two paths must be

$$|\tau_i - \tau_k| > \frac{1}{B} \qquad i, k = 1, 2, 3, \ldots, r \qquad (6.4)$$

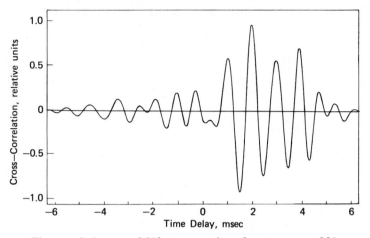

Time resolution $\tau_e = 0.012$ msec, number of averages $n_d = 256$

Figure 6.4   Cross-correlation function for multiple path acoustic experiment with $B = 800$ Hz and side reflection present.

to obtain clearly defined correlation peaks corresponding to the individual paths. In some cases, however, the analysis resolution can be improved using a unit impulse response calculation, as will be discussed later in Section 6.1.3.

6.1.2   *Cross-Spectra Measurements*

If the cross-correlation function identifies individual paths in a multiple path problem, then the same information must be available in its Fourier transform, the spectral density function. Specifically, taking the Fourier transform of Equation 6.2 yields

$$G_{xy}(f) = \sum_{k=1}^{r} H_k G_{xx}(f) e^{-j2\pi f \tau_k} \tag{6.5}$$

or in terms of complex polar notation,

$$G_{xy}(f) = |G_{xy}(f)| e^{-j\theta_{xy}(f)} \tag{6.6}$$

where

$$|G_{xy}(f)| = G_{xx}(f) \left[\sum_{i=1}^{r} \sum_{k=1}^{r} H_i H_k \cos 2\pi f (\tau_i - \tau_k) \right]^{1/2} \tag{6.7}$$

$$\theta_{xy}(f) = \tan^{-1} \left[\frac{\sum_{k=1}^{r} H_k \sin 2\pi f \tau_k}{\sum_{k=1}^{r} H_k \cos 2\pi f \tau_k} \right] \tag{6.8}$$

For a single path problem with a propagation time τ_1,

$$|G_{xy}(f)| = H_1 G_{xx}(f) \qquad \theta_{xy}(f) = 2\pi f \tau_1 \qquad (6.9)$$

Thus the phase angle reduces to a linear function of frequency as previously developed in Section 3.3.4. For a two path problem with propagation times τ_1 and τ_2,

$$|G_{xy}(f)| = G_{xx}(f)[H_1^2 + H_2^2 + 2H_1 H_2 \cos 2\pi f(\tau_2 - \tau_1)]^{1/2}$$

$$\theta_{xy}(f) = \tan^{-1}\left[\frac{H_1 \sin 2\pi f \tau_1 + H_2 \sin 2\pi f \tau_2}{H_1 \cos 2\pi f \tau_1 + H_2 \cos 2\pi f \tau_2}\right] \qquad (6.10)$$

From Equation 6.10, it is clear that the cross-spectrum magnitude for the two path problem will display an interference pattern having destructive interference notches equally spaced at frequency intervals of

$$\Delta f = \frac{1}{(\tau_2 - \tau_1)} \qquad (6.11)$$

with the first notch at $f = \Delta f/2$. From Equation 6.10, the cross-spectrum phase will be a ramp function with a frequency-dependent slope.

To find the coherence function between the input $x(t)$ and the output $y(t)$ in this problem, it is first necessary to solve for the autospectrum of $y(t)$ given in Equation 6.1. With the same arithmetic used to arrive at Equations 6.2 and 6.5, it follows that

$$R_{yy}(\tau) = \sum_{i=1}^{r} \sum_{k=1}^{r} H_i H_k R_{xx}(\tau + \tau_i - \tau_k) \qquad (6.12)$$

$$G_{yy}(f) = G_{xx}(f) \sum_{i=1}^{r} \sum_{k=1}^{r} H_i H_k \cos 2\pi f(\tau_i - \tau_k) \qquad (6.13)$$

From Equations 6.7 and 6.13, the coherence function between the input and output in a multiple path problem is then

$$\gamma_{xy}^2(f) = \frac{|G_{xy}(f)|^2}{G_{xx}(f)G_{yy}(f)} = 1 \qquad (6.14)$$

at all frequencies independent of the number of paths. This result is what one would expect. Assuming linear propagation and no extraneous noise, $y(t)$ is due solely to $x(t)$, hence it should be fully coherent with $x(t)$ no matter how many propagation paths are involved. It should be mentioned here that there are extreme multiple path situations, commonly called reverberation, which will cause the coherence between an input and output to be less than unity even for linear, noise-free propagation. These situations are discussed later in Section 7.3.4.

To illustrate the application of spectral measurements to a nondispersive multiple path problem, consider again the acoustic noise experiment in Figure 6.2 with a 3500 Hz bandwidth source. The coherence and phase computed between the input and output microphones with (a) no reflecting surfaces present and (b) the side reflection only are shown in Figure 6.5. For the direct-path case in Figure 6.5a, the phase is a ramp function closely corresponding to $\theta_{xy}(f) = 0.004\pi f$, as predicted by Equation 6.9 with a single propagation time $\tau_1 = 2.0$ msec. Furthermore, the coherence is almost exactly unity at all frequencies, as predicted by Equation 6.14, except below 200 Hz where the acoustic source was weak and masked by background noise.

For the two path case in Figure 6.5b, the phase is now a more complicated ramp function, as predicted by Equation 6.10, with $\tau_1 = 2.0$ msec and $\tau_2 = 3.9$ msec. The coherence, however, is still near unity at most frequencies above 200 Hz, as predicted by Equation 6.14. The notches in the coherence function at certain frequencies are due to the interference pattern in the cross-spectrum measurement, as predicted by Equation 6.11. Although the theoretical coherence is unity at all frequencies, the destructive interference notches sometimes drop the cross-spectrum measurement into the background noise causing a loss of coherence. This type of spurious notching of coherence functions is commonly observed and will be discussed further in Chapter 7.

A direct comparison of the spectral results in Figure 6.5 with the corresponding correlation results in Figure 6.3 quickly reveals why cross-correlation analysis is generally the more widely used analysis technique for applications involving nondispersive multiple path propagation problems. The same information is present in the cross-spectrum, but it is much more difficult to extract by visual inspection.

6.1.3   Unit Impulse Response Measurements

Referring back to Section 6.1.1, the identification of individual propagation paths in a cross-correlation analysis becomes increasingly difficult as the data bandwidth is reduced due to the spreading of the individual correlation peaks. In practice, if the input is artificial, the results can be optimized by simply making the bandwidth of the input as wide as feasible. However, the input of interest is more often a natural excitation that cannot be altered. Nevertheless, if the actual input spectrum extends over a wide range, but has concentrations of power in one or more narrow bands, then improved resolution of correlation peaks can often be achieved by using a unit impulse response calculation rather than direct cross-correlation analysis.

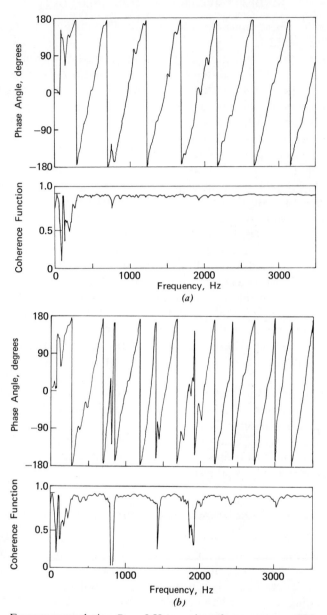

Frequency resolution $B_e = 8$ Hz, number of averages $n_d = 256$

Figure 6.5 Coherence function and phase angle for multiple path acoustic experiment with $B = 3500$ Hz. (*a*) Direct path only. (*b*) Side reflection present.

From Equation 1.45, the unit impulse response function $h(\tau)$ is given by the inverse Fourier transform of the frequency response function, that is,

$$h(\tau) = \int_{-\infty}^{\infty} H(f)e^{j2\pi f\tau} \, df \qquad (6.15)$$

From Equation 5.1, the optimum estimate of the frequency response function is given by

$$H(f) = \frac{G_{xy}(f)}{G_{xx}(f)} = \frac{S_{xy}(f)}{S_{xx}(f)} \qquad (6.16)$$

so it follows from Equation 6.15 that

$$h(\tau) = \int_{-\infty}^{\infty} \frac{S_{xy}(f)}{S_{xx}(f)} e^{j2\pi f\tau} \, df \qquad (6.17)$$

Now, from Equation 3.37, the cross-correlation function between $x(t)$ and $y(t)$ is given by

$$R_{xy}(\tau) = \int_{-\infty}^{\infty} S_{xy}(f)e^{j2\pi f\tau} \, df \qquad (6.18)$$

A direct comparison of Equations 6.17 and 6.18 shows that the unit impulse response function is like a cross-correlation function for an input $x(t)$ with a uniform spectral density of $S_{xx}(f) = 1$. In effect, the unit impulse response calculation produces the cross-correlation function for a uniform input spectral density. Hence one can computationally prewhiten the input data over its frequency range using a unit impulse response calculation, and sometimes improve the definition of individual propagation paths.

To illustrate this approach, consider the simple acoustical experiment outlined in Figure 6.6a. Random noise with a bandwidth of 10,000 Hz, but filtered to concentrate the power in a bandwidth of 450 Hz centered at 2000 Hz, is applied from the speaker. The autospectrum of the output signal is shown in Figure 6.6b. Now, the propagation time for the acoustic noise through its direct path to the microphone should be $\tau_1 = 0.61/c = 1.8$ msec, assuming a speed of sound of $c = 340$ meters/sec. Because of the back reflection, there is a second path where the propagation time should be $\tau_2 = 0.89/c = 2.6$ msec.

The cross-correlation and the unit impulse response functions measured between the speaker signal $x(t)$ and the microphone output $y(t)$ are shown in Figure 6.7. The form of the correlation function in Figure 6.7a is similar to that predicted for ideal narrow-band noise in Equation 3.65. However, the envelope of the correlation function gives no evidence of the two individual paths. On the other hand, the unit impulse response function in

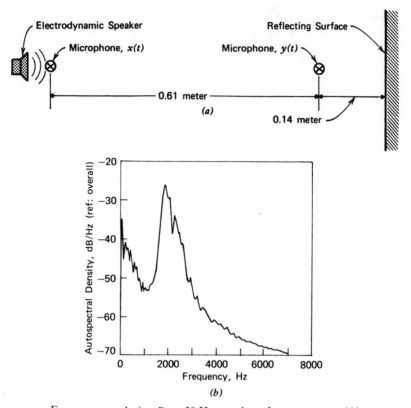

Frequency resolution $B_e = 30$ Hz, number of averages $n_d = 100$

Figure 6.6 Two path acoustic experiment. (*a*) Experimental setup. (*b*) Measured autospectrum of output.

Figure 6.7*b* clearly reveals the two paths. Furthermore, the envelopes of the two impulse response peaks reach maxima at the correct time delays of $\tau_1 = 1.8$ msec and $\tau_2 = 2.6$ msec.

The application of unit impulse response calculations to multiple path problems is often limited by a poor signal-to-noise ratio in the measured input signal $x(t)$. Specifically, if $x(t)$ is truly a narrow-band signal restricted to the frequency range $f_0 - B/2 \leq f \leq f_0 + B/2$, then the data outside this range will simply be extraneous noise and the unit impulse response calculation will provide no advantage over a direct cross-correlation calculation. To be effective, $x(t)$ must be well above the background noise over a much wider range than the apparent bandwidth of spectral peaks in the data. Furthermore, if the bandwidth of the output $y(t)$ is restricted by the frequency response function of the propagation path rather than by the bandwidth

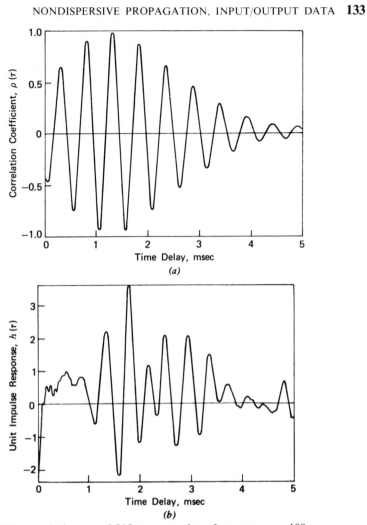

Time resolution $\tau_e = 0.015$ msec, number of averages $n_d = 100$

Figure 6.7   Results of two path acoustic experiment. (*a*) Cross-correlation function. (*b*) Impulse response function.

of the input signal, the unit impulse response calculation will again offer no advantage over cross-correlation analysis.

6.1.4   *Practical Considerations*

Beyond the various problems already discussed, there are numerous other considerations that limit the application of cross-correlation and cross-spectral analysis procedures to path identification problems. The most important of these are (a) substantial noise in the input and/or output measurements, and (b) scattering effects. The computations required to identify cross-correlation peaks corresponding to individual propagation paths increase dramatically as the number of paths and/or the extraneous measurement noise increases. Specifically, from the developments in Chapter 3, it is known that

$$\sum_{k=1}^{r} \rho_{xy}^2 \left(\frac{d_k}{c_k}\right) = 1 \qquad (6.19)$$

where $\rho_{xy}(d_k/c_k)$ is the correlation coefficient associated with each propagation path in a multiple-path problem. Hence the correlation coefficients associated with individual paths must diminish as the number of paths r increases. Beyond this, when extraneous noise is added to the input and/or output measurements with standard deviations of σ_m and σ_n, respectively, the correlation coefficient of each such peak is further diminished in accordance with dividing $\rho_{xy}(d_k/c_k)$ by the factor

$$\left[1 + \left(\frac{\sigma_n}{\sigma_y}\right)^2 + \left(\frac{\sigma_m}{\sigma_x}\right)^2 + \left(\frac{\sigma_n}{\sigma_y}\right)^2 \left(\frac{\sigma_m}{\sigma_x}\right)^2\right]^{1/2} \qquad (6.20)$$

Hence in problems involving several propagation paths and substantial amounts of input/output noise, one may be looking for correlation peaks with relatively small coefficients.

Again, from Chapter 3, the presence of extraneous noise in the input/output measurements will cause random errors in the correlation estimates. Specifically, from Table 3.2, the normalized random error in an estimate $\hat{R}_{xy}(\tau)$ is given by

$$\varepsilon[\hat{R}_{xy}(\tau)] = \left[\frac{1 + \rho_{xy}^{-2}(\tau)}{Nn_d}\right]^{1/2} \qquad (6.21)$$

where $\hat{R}_{xy}(\tau)$ is computed via the procedure of Equations 3.84 through 3.88. As a rule of thumb, if a correlation peak is to be detected, the value of the peak must be at least three standard deviations of the random error, that is,

$$\varepsilon[\hat{R}_{xy}(\tau)] \leq \tfrac{1}{3} \qquad (6.22)$$

Hence if a correlation peak with a coefficient of say $\rho_{xy}(\tau) = 0.10$ is to be detected, the total number of required data points Nn_d would be

$$Nn_d = \frac{[1 + \rho_{xy}^{-2}(\tau)]}{\varepsilon^2[\hat{R}_{xy}(\tau)]} \geq 909$$

Note here that n_d can be 1 if N is sufficiently large.

Referring back to Equation 6.2, this relationship for nondispersive multiple path propagation assumes that the gain factor of each path is uniform. This in turn produces correlation peaks that have the same shape as the auto-correlation function of the input $x(t)$. For the more general case of a non-uniform gain factor $|H_k(f)|$ for each path, by using $G_{xy}(f)$ and its complex conjugate $G_{xy}^*(f)$, Equation 6.18 becomes

$$R_{xy}(\tau) = \frac{1}{2} \int_0^\infty G_{xy}(f)e^{j2\pi f\tau}\, df + \frac{1}{2} \int_0^\infty G_{xy}^*(f)e^{-j2\pi f\tau}\, df \qquad (6.23)$$

and, in place of Equation 6.5,

$$G_{xy}(f) = \sum_{k=1}^{r} H_k(f)G_{xx}(f)e^{-j2\pi f\tau_k} \qquad (6.24)$$

Thus $G_{xy}(f)$ requires the complete $H_k(f)$ in both gain and phase. This shows that the cross-correlation peaks will now have a shape corresponding to an input $x(t)$ with a spectrum that would be approximated by $|H_k(f)|G_{xx}(f)$ rather than just $G_{xx}(f)$. Hence even if $x(t)$ is ideal white noise, the cross-correlation peaks may have a shape representing narrow-band noise if the path gain factor is narrow band in form. Nonuniform path gain factors can substantially widen correlation peaks and severely degradate the resolution of cross-correlation functions as applied to path identification problems. Note that the alternate analysis procedure of using unit impulse response calculations discussed in Section 6.1.3 will not resolve this problem. On the other hand, the fact that nonuniform path gain factors influence the shape of the correlation peaks can be used to estimate path frequency response functions, as will be discussed later in Section 6.4.

Concerning scattering, obstructions or a lack of homogeneity along a propagation path can cause wide angle reflections or an effective bending of the propagating energy flow. Such problems will generally blur and reduce the magnitude of peaks in the cross-correlation analysis, and might even cause correlation peaks to occur at spurious time delays. Any anomalies in a propagation path must be viewed with suspicion. For example, the propagation of acoustic noise in air can be significantly scattered by turbulence or gusting along the air path.

6.2 NONDISPERSIVE PROPAGATION, OUTPUT DATA ONLY

Situations sometimes arise where it is not possible to measure the input $x(t)$ in a multiple path propagation problem. In such cases, some information concerning multiple paths may still be extracted from analyses of the output $y(t)$ only.

6.2.1 *Autocorrelation Measurements*

Again consider a multiple path propagation problem as illustrated in Figure 6.1. Assuming nondispersive propagation through r number of paths, each with a uniform gain factor H_k, $i = 1, 2, 3, \ldots, r$, the output time history $y(t)$ will be as given by Equation 6.1. The autocorrelation of $y(t)$ is given in Equation 6.12, which may be rewritten as

$$R_{yy}(\tau) = R_{xx}(\tau) \sum_{k=1}^{r} H_k^2 + \sum_{\substack{i=1 \\ i \neq k}}^{r} \sum_{k=1}^{r} H_i H_k R_{xx}(\tau + \tau_i - \tau_k) \qquad (6.25)$$

In other words, the autocorrelation function of $y(t)$ will appear as a sum of autocorrelation functions of $x(t)$, each summed with a magnitude H_k^2, $k = 1, 2, 3, \ldots, r$, plus a collection of additional autocorrelation functions of $x(t)$, each offset by $\Delta \tau = (\tau_i - \tau_k)$ along the time scale, symmetrically about $\tau = 0$, and each with a magnitude $H_i H_k$ for $i \neq k = 1, 2, 3, \ldots, r$. Hence although no absolute values of propagation time are available from the results, differences in propagation times will be revealed.

To illustrate such results, consider again the three-path acoustic noise experiment outlined previously in Figure 6.2. The autocorrelation functions of the output $y(t)$ computed for various reflection configurations with an input bandwidth of $B = 8000$ Hz are shown in Figure 6.8. From Equation 6.25, the autocorrelation function for the direct path case should be simply

$$R_{yy}(\tau) = H_1^2 R_{xx}(\tau) \qquad (6.26)$$

exactly as computed in Figure 6.8*a* with $H_1 = 1$. With the side reflection alone, the result should be

$$R_{yy}(\tau) = (H_1^2 + H_2^2)R_{xx}(\tau) + H_1 H_2 R_{xx}(\tau + \tau_1 - \tau_2) \\ + H_2 H_1 R_{xx}(\tau + \tau_2 - \tau_1) \qquad (6.27)$$

where $\tau_1 - \tau_2 = -1.9$ msec and $\tau_2 - \tau_1 = 1.9$ msec. This expected result is clearly displayed in Figure 6.8*b*. For the back reflection alone, the expected result is

$$R_{yy}(\tau) = (H_1^2 + H_3^2)R_{xx}(\tau) + H_1 H_3 R_{xx}(\tau + \tau_1 - \tau_3) \\ + H_3 H_1 R_{xx}(\tau + \tau_3 - \tau_1) \qquad (6.28)$$

where $\tau_1 - \tau_3 = -3$ msec and $\tau_3 - \tau_1 = 3$ msec. Again, this is consistent with the measured result in Figure 6.8*c*. Finally, with both reflecting surfaces present, Equation 6.25 indicates that the autocorrelation function of $y(t)$ should involve seven terms, namely

$$
\begin{aligned}
R_{yy}(\tau) = (H_1^2 + H_2^2 + H_3^2)R_{xx}(\tau) + H_1 H_2 R_{xx}(\tau + \tau_1 - \tau_2) \\
+ H_2 H_1 R_{xx}(\tau + \tau_2 - \tau_1) + H_1 H_3 R_{xx}(\tau + \tau_1 - \tau_3) \\
+ H_3 H_1 R_{xx}(\tau + \tau_3 - \tau_1) + H_2 H_3 R_{xx}(\tau + \tau_2 - \tau_3) \\
+ H_3 H_2 R_{xx}(\tau + \tau_3 - \tau_2)
\end{aligned} \tag{6.29}
$$

where

$\tau_1 - \tau_2 = -1.9$ msec	$\tau_1 - \tau_3 = -3.0$ msec	$\tau_2 - \tau_3 = -1.1$ msec
$\tau_2 - \tau_1 = 1.9$ msec	$\tau_3 - \tau_1 = 3.0$ msec	$\tau_3 - \tau_2 = 1.1$ msec

From Figure 6.8d, the side peaks at $\tau = \pm 1.9$ msec are clearly apparent, but the side peaks at $\tau = \pm 3.0$ msec are barely discernable and the peaks at $\tau = \pm 1.1$ msec are completely obscured. The problem, of course, is that the bandwidth of the data is not sufficiently wide to permit a proper resolution of all the propagation time differences involved in this experiment. The resolution problem here is even more severe than in the corresponding cross-correlation results shown previously in Figure 6.3d because the cross-correlation analysis does not generate peaks at time delay differences for the reflected paths; that is, there are no peaks in the cross-correlation results at $\tau = \pm |\tau_2 - \tau_3|$.

6.2.2   *Autospectra Measurements*

Now consider the autospectral density of the output $y(t)$ in a nondispersive propagation problem as illustrated in Figure 6.1. The autospectrum of $y(t)$ is given by the Fourier transform of the autocorrelation function, as defined previously in Equation 6.13, which may be rewritten as

$$
G_{yy}(f) = G_{xx}(f) \left[\sum_{k=1}^{r} H_k^2 + \sum_{\substack{i=1 \\ i \neq k}}^{r} \sum_{k=1}^{r} H_i H_k \cos 2\pi f (\tau_i - \tau_k) \right] \tag{6.30}
$$

This result shows that the autospectrum of $y(t)$ will display destructive interference patterns with the various interference notches spaced in frequency at intervals of

$$
\Delta f_{ik} = \frac{1}{|\tau_i - \tau_k|} \qquad i \neq k = 1, 2, 3, \ldots, r \tag{6.31}
$$

Again, using results from the three path acoustic noise experiment outlined in Figure 6.2, the autospectra of the output $y(t)$ computed for various

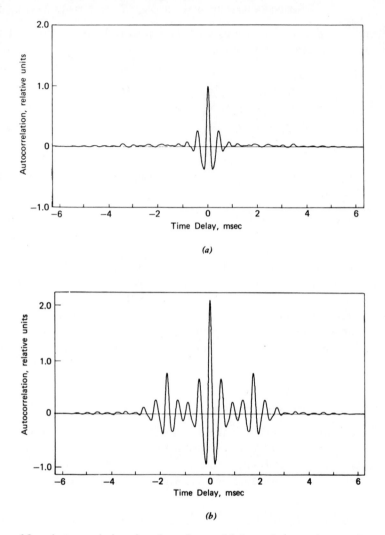

Figure 6.8 Autocorrelation functions for multiple path acoustic experiment in Figure 6.2 with $B = 8000$ Hz. (a) Direct path only. (b) Side reflection present.

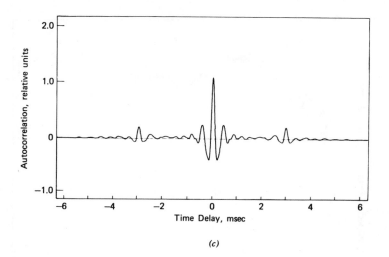

(c)

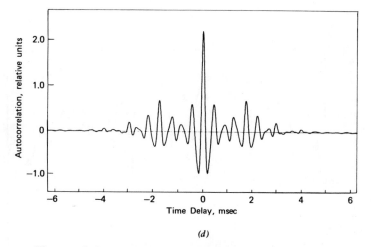

(d)

Time resolution $\tau_e = 0.012$ msec, number of averages $n_d = 256$

Figure 6.8 Autocorrelation functions for multiple path acoustic experiment in Figure 6.2 with $B = 8000$ Hz. (c) Back reflection present. (d) Side and back reflections present.

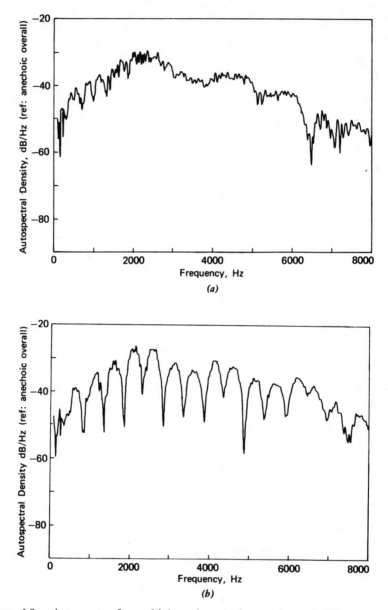

Figure 6.9 Autospectra for multiple path acoustic experiment in Figure 6.2 with $B = 8000$ Hz. (*a*) Direct path only. (*b*) Side reflection present.

140

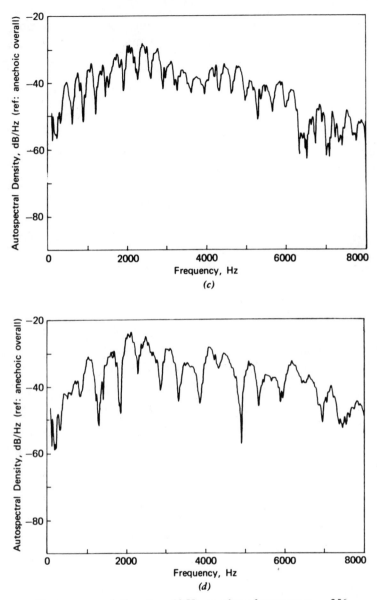

(c)

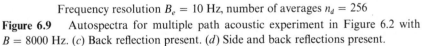

(d)

Frequency resolution $B_e = 10$ Hz, number of averages $n_d = 256$

Figure 6.9 Autospectra for multiple path acoustic experiment in Figure 6.2 with $B = 8000$ Hz. (c) Back reflection present. (d) Side and back reflections present.

reflection configurations are as shown in Figure 6.9. From Equation 6.30, the autospectrum of $y(t)$ with no reflections is simply

$$G_{yy}(f) = H_1^2 G_{xx}(f) \tag{6.32}$$

as shown in Figure 6.9a. When the side reflection alone is added, Equation 6.30 gives

$$G_{yy}(f) = G_{xx}(f)[H_1^2 + H_2^2 + 2H_1 H_2 \cos 2\pi f(\tau_2 - \tau_1)] \tag{6.33}$$

which is the sum of the spectra through the two paths with a superimposed interference pattern at the frequency

$$\Delta f_{12} = \frac{1}{\tau_2 - \tau_1} \approx 530 \text{ Hz}$$

When the back reflection alone is added, the result is the same as in Equation 6.33 with H_2 and τ_2 replaced by H_3 and τ_3, respectively, that is, the interference notches now have a frequency separation of

$$\Delta f_{13} = \frac{1}{\tau_3 - \tau_1} \approx 330 \text{ Hz}$$

These predicted results are clearly apparent in the measured spectra shown in Figures 6.9b and c, respectively. Finally, with both reflecting surfaces present, Equation 6.30 yields

$$\begin{aligned} G_{yy}(f) = G_{xx}(f)[H_1^2 + H_2^2 + H_3^2 + 2H_1 H_2 \cos 2\pi f(\tau_2 - \tau_1) \\ + 2H_1 H_3 \cos 2\pi f(\tau_3 - \tau_1) + 2H_2 H_3 \cos 2\pi f(\tau_3 - \tau_2)] \end{aligned} \tag{6.34}$$

This result suggests an output spectrum that is the sum of the spectra through the three paths with a superimposed interference pattern including three components at the frequencies

$$\Delta f_{12} \approx 530 \text{ Hz} \qquad \Delta f_{13} \approx 330 \text{ Hz} \qquad \Delta f_{23} \approx 910 \text{ Hz}$$

The measured results for this case in Figure 6.9d reveal the presence of a complex interference pattern, but the individual components identifying specific propagation paths are no longer discernible.

6.2.3   Practical Considerations

The various problems discussed for the cross-correlation and cross-spectra analysis procedures in Section 6.1.4 apply to the autocorrelation and autospectra measurements as well, except now the extraneous noise problem is far more severe. In a cross-correlation analysis, extraneous noise at the input or output reduces the relative contribution of individual correlation peaks and increases the random error in the analysis, but it does not distort or bias the results. In an autocorrelation analysis, however, the noise $n(t)$ at the output adds directly to the output signal $v(t)$ of interest, that is,

$$y(t) = v(t) + n(t) \qquad R_{yy}(\tau) = R_{vv}(\tau) + R_{nn}(\tau) \qquad (6.35)$$

Hence noise in $y(t)$ biases $R_{yy}(\tau)$ and may obscure the peaks corresponding to propagation time delay differences. The same conclusions apply to the autospectra measurements.

Even when no extraneous noise is present in the output measurements, autocorrelation and autospectra analysis offer a very poor path identification tool. Either calculation may identify the presence of two paths and the difference in their propagation times from the source to the output measurement. When three or more paths are present, however, the identification of individual paths becomes more difficult. Of course, specialized analyses can sometimes be used to augment the detection of propagation time differences. For example, a Fourier transform of the output spectrum (called the Cepstrum) will convert the individual interference components into more readily identifiable spectral peaks at frequencies corresponding to $\Delta f_{ik} = 1/|\tau_i - \tau_k|$; $i \neq k = 1, 2, 3, \ldots, r$.

For cases where the energy through each path approaches the output location from a different direction, as commonly occurs in acoustic noise and electromagnetic radiation problems, individual paths can often be isolated by beam-forming operations in the output measurement, namely, by using an array or antenna for the output measurement. In other cases where the time delay differences due to possible multiple path propagation can be anticipated, curve fitting operations based on conventional regression analysis procedures might prove helpful. Further discussions of related problems are presented in Chapter 7.

6.3 DISPERSIVE PROPAGATION

Consider next the situation where the propagation paths in Figure 6.1 are frequency dispersive, that is, the propagation velocity is a function of frequency. In practice, dispersion generally involves a propagation velocity that increases with frequency. Hence given an input $x(t)$ that propagates through a frequency dispersive medium, the higher frequency components of $x(t)$ will arrive at an output measurement $y(t)$ sooner than the lower frequency components. The apparent propagation speed of the waves at a given frequency is called the *group velocity* c_g, which is related but not equal to the phase velocity c_p of the waves. Specifically, assuming $c_p \sim \lambda^n$ where λ

is wavelength, $c_g = (1 - n)c_p$ [6.1]. For example, the phase velocity of flexural waves in a long thin beam is given by [6.2]

$$c_p = \sqrt{1.8hfc_l} \sim \frac{1}{\lambda} \tag{6.36}$$

where h is the thickness of the beam and c_l is the longitudinal wave velocity (speed of sound) in the beam. Hence $n = -1$ in this case and the group velocity of the flexural waves will be

$$c_g = 2c_p = 2\sqrt{1.8hfc_l} \tag{6.37}$$

Note from Equation 6.37 that $c_g \sim f^{1/2}$.

6.3.1 Single Path Problems

Even for a single propagation path between an input $x(t)$ and an output $y(t)$, the broad-band cross-correlation analysis procedures discussed in Section 6.1.1 will generally not provide meaningful results if the propagation is dispersive. Since the waves at different frequencies travel at different velocities, no distinct peak will be observed in the cross-correlation plot. It is obvious, however, that this difficulty can be suppressed by simply analyzing the data in contiguous frequency bands, each sufficiently narrow to limit the group velocities of the waves within the bands to a range that will provide an acceptably well-resolved correlation peak. For the case of flexural waves in beams and plates where $c_g \sim f^{1/2}$, a bandwidth where the upper frequency limit is twice the lower limit (one octave) will often give acceptable results.

To illustrate the above points, consider the simple experiment performed by White [6.3], as outlined in Figure 6.10. A long, straight aluminum beam with a thickness of $h = 0.0032$ meter was subjected to random excitation at one end. Two accelerometers were mounted on the span with a separation distance of $d = 1.2$ meters. The ends of the beam were heavily damped to suppress end reflections. The cross-correlation functions computed between the two accelerometers in three individual octave band frequency intervals centered at 1, 2, and 8 kHz are shown in Figure 6.11. From Equation 6.37,

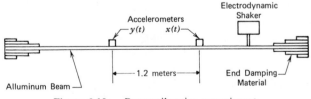

Figure 6.10   Beam vibration experiment.

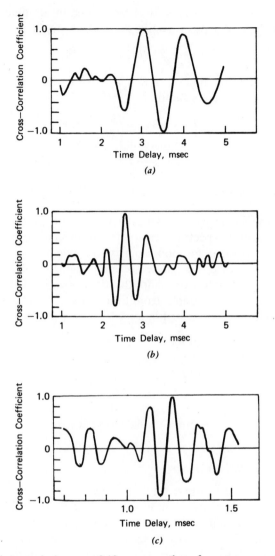

Time resolution $\tau_e = 0.10$ msec, number of averages $n_d = 15$

Figure 6.11 Cross-correlation functions in octave frequency bands for beam vibration experiment. (*a*) 1000 Hz. (*b*) 2000 Hz. (*c*) 8000 Hz. (The data in Figures 6.11 and 6.13 are taken from Reference 6.3 with the permission of The Acoustical Society of America and the author.

145

with $h = 0.0032$ meter and $c_l = 5150$ meters/sec in aluminum, the group velocity of the flexural waves traveling down the beam is

$$c_g = 2\sqrt{1.8hfc_l} = 11f^{1/2} \tag{6.38}$$

Hence the propagation times $\tau = (d/c_g)$ between the two accelerometers separated by $d = 1.2$ meters should be $\tau = 0.11f^{-1/2}$ or

$$\tau = \begin{cases} 3.5 \text{ msec at } f = 1000 \text{ Hz} \\ 2.5 \text{ msec at } f = 2000 \text{ Hz} \\ 1.2 \text{ msec at } f = 8000 \text{ Hz} \end{cases}$$

These predicted time delays are fully consistent with the measured results in Figure 6.11.

Of course, the single path dispersive propagation problem can also be evaluated using cross-spectral density functions (coherence and phase), as previously illustrated for nondispersive propagation in Section 6.1.2. For the dispersive case, the phase factor will no longer be linear with frequency, but instead will have a shape that defines the frequency dependence of the propagation. For example, from Equations 3.74 and 6.37, the flexural wave experiment illustrated in Figure 6.10 would produce a phase angle between $x(t)$ and $y(t)$ of

$$\theta(f) = \frac{2\pi f d}{c_g} = \frac{\pi d f^{1/2}}{(1.8hc_l)^{1/2}} \tag{6.39}$$

Hence the flexural waves in a beam would be indicated in a cross-spectrum analysis by a phase that is proportional to the square root of frequency.

6.3.2   Multiple Path Problems

Now consider the case where there is dispersive propagation through two or more parallel paths between an input $x(t)$ and an output $y(t)$, as shown in Figure 6.1. Conceptually, this problem can be evaluated using a band-limited cross-correlation analysis to suppress dispersion effects, as discussed in Section 6.3.1. However, two conflicting requirements now arise: (a) the need for a narrow bandwidth analysis to suppress dispersion and (b) the need for a wide bandwidth analysis to suppress overlapping of adjacent correlation peaks. In many cases, these two conflicting requirements will prevent the acquisition of meaningful results. However, in some cases, an acceptable compromise can be reached.

As an illustration, consider the two path experiment performed by White [6.3], as outlined in Figure 6.12. A long, straight aluminum beam with a

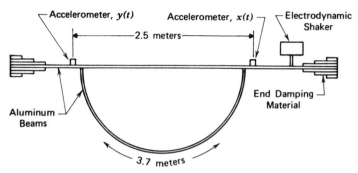

Figure 6.12   Two path beam vibration experiment.

thickness of $h = 0.0032$ meter was welded to a semicircular beam of similar thickness, and random excitation was applied at one end. Two accelerometers were mounted on the span of the straight beam with a separation distance of $d_1 = 2.5$ meters along the straight path and $d_2 = 3.7$ meters along the semicircular path. The ends of the straight beam were heavily damped to suppress end reflections. A typical cross-correlation function computed between the two accelerometers in an octave band centered at 850 Hz is shown in Figure 6.13. From Equation 6.38, the group velocity of the flexural waves traveling down the two beam paths is

$$c_g = 11f^{1/2} = 320 \text{ meters/sec}$$

Hence the propagation times between the two accelerometers through the straight and semicircular paths should be

$$\tau_1 = 7.8 \text{ msec} \qquad \tau_2 = 12 \text{ msec}$$

Referring to Figure 6.13, the envelope of the measured cross-correlation function clearly reveals two peaks at these predicted time delays.

The various practical considerations discussed in Section 6.1.4 apply to dispersive propagation problems as well. In particular, the measurement noise problem detailed in Equations 6.19 through 6.22 would apply directly if the dispersion were fully suppressed by the band-limited analysis operations. However, because of the conflicting bandwidth requirements, dispersion is usually not fully suppressed and overlapping of correlation peaks is common in dispersive multiple path problems. This tends to make the correlation coefficients of individual peaks even smaller than predicted by Equation 6.20 and thus the necessary amount of averaging even greater than suggested by Equations 6.21 and 6.22.

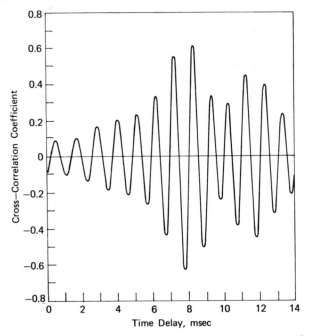

Time resolution $\tau_e = 0.10$ msec, number of averages $n_d = 15$

Figure 6.13   Cross-correlation function in octave band at 850 Hz for two path beam vibration experiment.

6.3.3   *Mixed Path Problems*

Situations often arise where the energy of concern propagates through a nonhomogeneous path with two or more propagation velocities. For mechanical structures, the propagation of energy may further occur in the form of both nondispersive longitudinal waves and dispersive flexural waves. A common example is the propagation of high-frequency vibration (of concern in noise control engineering) through large steel-reinforced concrete structures such as multiple-story commercial buildings. Dynamic energy can propagate through a steel-reinforced concrete member as (*a*) longitudinal waves through the concrete, (*b*) longitudinal waves through the reinforcing rods, and/or (*c*) flexural waves down the entire member. Cross-correlation analyses in octave bands at various center frequencies will often identify these three individual contributions. The longitudinal waves in particular can be readily distinguished from flexural waves because they are nondispersive and thus appear with the same time delay at all frequencies.

As an example, consider the vibration experiment performed on the ground floor of a multiple-story office building illustrated in Figure 6.14. An electro-

Steel Reinforced Concrete Floor

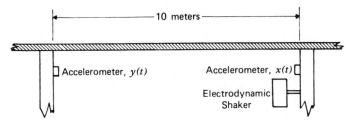

Figure 6.14 Vibration experiment in steel-reinforced concrete office building.

dynamic shaker was attached to a column supporting a steel-reinforced floor structure. One accelerometer was mounted near the shaker and a second accelerometer was located on another supporting column. The distance between the two accelerometers through the floor slab was about 10 meters and the slab thickness was about 0.37 meter including a 0.1 meter topping slab. Random vibration was applied from the shaker in octave frequency bands centered at 500 and 1000 Hz. The cross-correlation results are shown in Figure 6.15. The analysis was performed with a time delay resolution of $\tau_e = 0.05$ msec and $n_d = 20$ averages.

The results in Figure 6.15 reveal two correlation peaks that occur at the same delay times in both the 500 Hz and 1000 Hz octave bands, namely at

$$\tau_1 = 2 \text{ msec} \qquad \tau_2 = 6 \text{ msec}$$

The dominant peak at $\tau_1 = 2$ msec corresponds to a propagation velocity of 5000 meters/sec, the approximate speed of sound in steel. It follows that longitudinal waves in the steel reinforcing rods of the floor were the primary vehicle for energy transmission between the two measurement points at these frequencies. The peak at $\tau_2 = 6$ msec, clearly apparent at 500 Hz and less obvious but discernible at 1000 Hz, corresponds to a propagation velocity of 1700 meters/sec. This is the approximate speed of sound in lightweight concrete [6.2] and hence probably represents longitudinal wave propagation through the concrete material. The energy transmitted through the concrete is substantially less than through the reinforcing rods because concrete has a much higher loss factor than steel.

Concerning flexural waves, the entire floor structure is too thick for classical flexural wave response at frequencies above 500 Hz [6.2]. However, individual elements of the floor, such as the topping slab or the stiffeners, might transmit flexural waves at the frequencies of interest. The data in Figure 6.15a reveal a well-defined correlation peak at $\tau_3 = 13$ msec in the octave band centered at 500 Hz. From Equation 6.37, this is the correct time delay

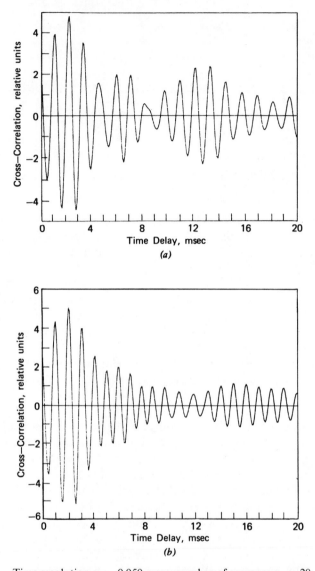

Time resolution $\tau_e = 0.050$ msec, number of averages $n_d = 20$

Figure 6.15 Cross-correlation function in octave frequency bands for office building vibration experiment. (*a*) 500 Hz. (*b*) 1000 Hz. (These data were acquired with the assistance of Müller-BBM GmbH, Munich, Germany.)

150

for a flexural wave propagating through some element of the floor structure with an effective thickness of 0.1 meter, which corresponds to the thickness of the topping slab. On the other hand, the peak at 13 msec might also represent another longitudinal wave path through a different section of the building structure.

6.4 ESTIMATION OF PATH CHARACTERISTICS

Sections 6.1 through 6.3 address the general problem of identifying the possible propagation paths between single input and output measurements. It is assumed in these developments that each propagation path has a frequency response function of $H(f) = H$, a constant. In practice, of course, $H(f)$ might be other than a constant and its estimated characteristics may be of central interest.

6.4.1 *Path Frequency Response Estimates*

Referring back to Equation 6.2, if the frequency response function of a propagation path is some arbitrary $H(f)$ rather than a constant H, then the cross-correlation function for the multiple path case is given by Equation 6.23 rather than Equation 6.2. Now, the cross-correlation peaks identifying each path will have a shape corresponding to an input $x(t)$ with a spectrum of the form $|H_k(f)|G_{xx}(f)$, $k = 1, 2, 3, \ldots, r$, rather than $G_{xx}(f)$ alone. As noted in Section 6.1.4, this can pose a problem in the determination of individual paths since nonuniform frequency response functions will generally tend to spread the correlation peaks and aggravate the overlap problem. On the other hand, if the correlation peaks completely separate along the time delay axis, information concerning the frequency response functions of individual paths can be extracted from the cross-correlation results.

Specifically, consider a cross-correlation function computed for a multiple path propagation problem as shown in Figure 6.16. From Equations 6.23 and 6.24, the isolated cross-correlation function of the results for the kth path alone is given by $R_{xy_k}(\tau)$ instead of by $R_{xy}(\tau)$, that is,

$$R_{xy_k}(\tau) = \frac{1}{2} \int_0^\infty G_{xy_k}(f)e^{j2\pi f\tau} \, df + \frac{1}{2} \int_0^\infty G_{xy_k}^*(f)e^{-j2\pi f\tau} \, df \qquad (6.40)$$

where

$$G_{xy_k}(f) = H_k(f)G_{xx}(f)e^{-j2\pi f\tau_k} \qquad (6.41)$$

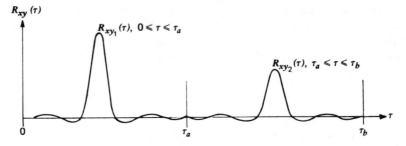

Figure 6.16   Separable cross-correlation function for a multiple path problem.

Hence assuming no noise in the input measurement $x(t)$, the frequency response function of the kth path is estimated from Equation 6.41 by

$$H_k(f) = \frac{G_{xy_k}(f)}{G_{xx}(f)} \, e^{j2\pi f \tau_k} \qquad (6.42)$$

Note in Equation 6.42 that $H_k(f)$ may have a phase factor different from zero in addition to the phase shift associated with the propagation time τ_k. If the phase factor is not a linear function of frequency, then the path is dispersive. In any case, the two phase components will be combined and not individually identified, that is,

$$H_k(f) = |H_k(f)| e^{-j[\theta_k(f) - 2\pi f \tau_k]} \qquad (6.43)$$

Finally, it should be mentioned that the procedure for estimating individual path frequency response functions outlined in Equations 6.40 through 6.42 is adversely impacted by even modest truncation errors in the correlation function describing each path. This means that the procedure will generally not provide accurate results if there is significant overlapping of correlation peaks. Furthermore, if the procedure is applied to dispersive path problems, it is important that the filters used to band limit the input/output data, as required to suppress dispersion effects, be carefully matched to have identical phase characteristics. Otherwise, the relative phase shift of the band-limiting filters will appear in the phase factor of $H_k(f)$ given by Equation 6.43.

Other possible bias errors as well as the random errors in the frequency response estimates given by Equation 6.42 are discussed in Chapters 5 and 11.

6.4.2   *Applications Using Transients*

All of the applications of correlation and spectral analysis to path identification problems discussed in this section could be accomplished just as well using transient rather than stationary random inputs. Specifically, from

Equation 6.1, let $x(t)$ be a short duration transient such that its duration Δt is

$$\Delta t < (\tau_i - \tau_k) \qquad i, k = 1, 2, 3, \ldots, r \qquad (6.44)$$

The output measurement given by

$$y(t) = \sum_{k=1}^{r} H_k x(t - \tau_k) \qquad (6.45)$$

will then be a series of transients, each occurring at a time displacement τ_k, $k = 1, 2, 3, \ldots, r$, corresponding to the delay time identifying the kth propagation path.

From Section 4.1.2, all of the spectral relationships developed in this chapter apply where the energy spectra $\mathscr{G}_{xy}(f)$ and $\mathscr{G}_{yy}(f)$ replace the stationary spectra $G_{xy}(f)$ and $G_{yy}(f)$. For example, the path separation criterion of Equation 6.4 applies where B is the bandwidth of the energy spectrum $\mathscr{G}_{yy}(f)$. Similarly, the frequency response estimation procedures of Equations 6.41 and 6.42 apply using energy spectra in place of the stationary spectra.

The primary difference between the use of transients as opposed to stationary random inputs is in the way measurement noise is suppressed. For a cross-correlation or cross-spectrum analysis, the accuracy of the estimates is improved by increasing the number of averages n_d as given by Equations 3.89 and 3.90, which means simply increasing the total length of the experiment. For transient inputs, the averaging must be done over a collection of transients, that is, the experiment must be repeated many times. In those cases where the transient is deterministic and can be repeated exactly with no measurement noise at the input, the averaging can be accomplished on the output signals only.

REFERENCES

6.1 Rayleigh, J. W. S., *The Theory of Sound*, Vol. 1, Appendix, Dover Publications, New York, 1945.

6.2 Cremer, L., Heckl, M., and Ungar, E. E., *Structure-Borne Sound*, Springer-Verlag, New York, 1973.

6.3 White, P. H., "Cross-Correlation In Structural Systems: Dispersive and Non-dispersive Waves," *Journal of the Acoustical Society of America*, Vol. 45, No. 5, p. 1118, May 1969.

CHAPTER 7

SINGLE INPUT/MULTIPLE OUTPUT PROBLEMS

Of interest now are problems involving a single input, measured or unmeasured, that produces several output measurements. Each output may be an individual measurement or may represent the output of an array of measurements used for beam-forming purposes. In either case, it is assumed that there is a constant parameter linear system between the input and each measured output, and that all unknown deviations from this ideal case can be included in uncorrelated extraneous noise terms at the output.

7.1 CORRELATION AND SPECTRA RELATIONSHIPS

Consider the system shown in Figure 7.1 consisting of a single stationary (ergodic) random input $x(t)$ producing r measured outputs $y_i(t)$, $i = 1, 2, \ldots, r$. From Equation 4.1, each output is given by

$$y_i(t) = \int_0^\infty h_i(\tau)x(t - \tau)\, d\tau + n_i(t) \qquad i = 1, 2, \ldots, r \qquad (7.1)$$

Taking finite Fourier transforms over a long record length T yields

$$Y_i(f, T) = H_i(f)X(f, T) + N_i(f, T) \qquad i = 1, 2, \ldots, r \qquad (7.2)$$

From the assumption of stationarity,

$$\mu_{y_i} = E[y_i(t)] = \mu_x \int_0^\infty h_i(\tau)\, d\tau + \mu_{n_i} \qquad i = 1, 2, \ldots, r \qquad (7.3)$$

Thus for all $y_i(t)$, $\mu_y = 0$ when both $\mu_x = 0$ and $\mu_n = 0$. Such conditions will be assumed for convenience to simplify later calculations and interpretations. It will also be assumed that $x(t)$ and $n_i(t)$ are uncorrelated, that is, $E[x(t)n_i(t)] = 0$ for all i.

7.1.1   *General Input/Output Relationships*

The autocorrelation function for any of the outputs $y_i(t)$ may be derived, as in Equation 4.3, to be

$$R_{y_iy_i}(\tau) = \int_0^\infty \int_0^\infty h_i(\alpha)h_i(\beta)R_{xx}(\tau + \alpha - \beta)\,d\alpha\,d\beta + R_{n_in_i}(\tau) \qquad (7.4)$$

Similarly, the autospectral density function is given from Equation 4.8 by

$$G_{y_iy_i}(f) = |H_i(f)|^2 G_{xx}(f) + G_{n_in_i}(f) \qquad (7.5)$$

The cross-correlation function between the input $x(t)$ and any of the outputs $y_i(t)$ is given by Equation 4.5 as

$$R_{xy_i}(\tau) = \int_0^\infty h_i(\alpha)R_{xx}(\tau - \alpha)\,d\alpha \qquad (7.6)$$

In the frequency domain, the corresponding cross-spectral density function relation from Equation 4.9 is

$$G_{xy_i}(f) = H_i(f)G_{xx}(f) \qquad (7.7)$$

Equations 7.1 through 7.7 are exactly the same as for single input/single output problems with extraneous output noise.

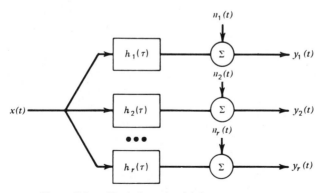

Figure 7.1   Single input/multiple output system.

Consider now the cross-correlation function between any two of the output records $y_i(t)$ and $y_j(t)$ where $i \neq j$. This is given by

$$R_{y_i y_j}(\tau) = E[y_i(t)y_j(t + \tau)]$$

$$= \int_0^\infty \int_0^\infty h_i(\alpha)h_j(\beta)E[x(t - \alpha)x(t + \tau - \beta)] \, d\alpha \, d\beta$$

$$= \int_0^\infty \int_0^\infty h_i(\alpha)h_j(\beta)R_{xx}(\tau + \alpha - \beta) \, d\alpha \, d\beta \tag{7.8}$$

where all other terms average to zero. This complicated result is difficult to compute and to interpret. On the other hand, by taking the Fourier transform of Equation 7.8, the cross-spectral density function between any two of the output records is given by the simple expression

$$G_{y_i y_j}(f) = H_i^*(f)H_j(f)G_{xx}(f) \tag{7.9}$$

This result shows that the measurement of $G_{y_i y_j}(f)$ plus a separate knowledge of $H_i(f)$ and $H_j(f)$ enable one to estimate $G_{xx}(f)$ when $G_{xx}(f)$ cannot be measured directly.

From Equations 7.5 and 7.9, the coherence function between any two of the output records is given by

$$\gamma_{y_i y_j}^2(f) = \frac{|G_{y_i y_j}(f)|^2}{G_{y_i y_i}(f)G_{y_j y_j}(f)}$$

$$= \frac{|H_i(f)|^2|H_j(f)|^2 G_{xx}^2(f)}{[|H_i(f)|^2 G_{xx}(f) + G_{n_i n_i}(f)][|H_j(f)|^2 G_{xx}(f) + G_{n_j n_j}(f)]} \tag{7.10}$$

Various special cases of Equation 7.10 are of physical interest. First, assume that $G_{n_i n_i}(f) = G_{n_j n_j}(f) = 0$. It follows that

$$\gamma_{y_i y_j}^2(f) = 1 \tag{7.11}$$

which indicates that $y_i(f)$ and $y_j(t)$ come from a common source $x(t)$ uncorrupted by extraneous noise. Next assume that $H_i(f) = H_j(f) = H(f)$ and let

$$G_{vv}(f) = |H(f)|^2 G_{xx}(f) \tag{7.12}$$

It now follows that

$$\gamma_{y_i y_j}^2(f) = \frac{G_{vv}^2(f)}{[G_{vv}(f) + G_{n_i n_i}(f)][G_{vv}(f) + G_{n_j n_j}(f)]} \tag{7.13}$$

which demonstrates that the coherence function will always be less than unity if there is extraneous noise in either output measurement. Finally, assume that $H_i(f) = H(f)$ and $H_j(f) = 1$. For this case,

$$\gamma_{y_i y_j}^2(f) = \frac{|H(f)|^2 G_{xx}^2(f)}{[|H(f)|^2 G_{xx}(f) + G_{n_i n_i}(f)][G_{xx}(f) + G_{n_j n_j}(f)]} \quad (7.14)$$

The last two cases can be studied further using specific forms for $H(f)$, as pursued in Reference 7.1 where many other related matters are discussed.

7.1.2 Single Input/Two Output Relationships

Consider the special case of a single input/two output system as illustrated in Figure 7.2. From the relationships in Section 7.1.1, the following frequency domain equations apply to this situation, assuming that the noise terms $n_1(t)$ and $n_2(t)$ are uncorrelated with each other and with the input $x(t)$. The dependence on frequency f is omitted here to simplify the notation.

$$\begin{aligned}
G_{xn_1} &= G_{un_1} = G_{xn_2} = G_{vn_2} = G_{n_1 n_2} = 0 \\
G_{y_1 y_1} &= G_{uu} + G_{n_1 n_1} = |H_1|^2 G_{xx} + G_{n_1 n_1} \\
G_{y_2 y_2} &= G_{vv} + G_{n_2 n_2} = |H_2|^2 G_{xx} + G_{n_2 n_2} \\
G_{xy_1} &= G_{xu} = H_1 G_{xx} \qquad G_{xy_2} = G_{xv} = H_2 G_{xx} \\
G_{y_1 y_2} &= G_{uv} = H_1^* H_2 G_{xx}
\end{aligned} \quad (7.15)$$

Also for this model, the coherence function between the output records is given by

$$\gamma_{y_1 y_2}^2 = \frac{|G_{y_1 y_2}|^2}{G_{y_1 y_1} G_{y_2 y_2}} = \frac{|G_{uv}|^2}{G_{y_1 y_2} G_{y_2 y_2}} = \gamma_{xy_1}^2 \gamma_{xy_2}^2 \quad (7.16)$$

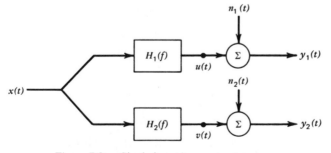

Figure 7.2 Single input/two output system.

The above result follows from the fact that

$$|G_{uv}|^2 = |H_1^* H_2 G_{xx}|^2 = (|H_1|^2 G_{xx})(|H_2|^2 G_{xx}) = G_{uu} G_{vv}$$

$$\gamma_{xy_1}^2 = \frac{|G_{xy_1}|^2}{G_{xx} G_{y_1 y_1}} = \frac{|H_1 G_{xx}|^2}{G_{xx} G_{y_1 y_1}} = \frac{G_{uu}}{G_{y_1 y_1}} \tag{7.17}$$

$$\gamma_{xy_2}^2 = \frac{|G_{xy_2}|^2}{G_{xx} G_{y_2 y_2}} = \frac{|H_2 G_{xx}|^2}{G_{xx} G_{y_2 y_2}} = \frac{G_{vv}}{G_{y_2 y_2}}$$

Two special cases will now be considered. First, assume that $n_1(t) = 0$ so that $G_{n_1 n_1} = 0$. It follows that

$$\gamma_{y_1 y_2}^2 = \gamma_{u y_2}^2 = \gamma_{x y_2}^2 = \frac{G_{vv}}{G_{y_2 y_2}} \tag{7.18}$$

Thus,

$$G_{vv} = \gamma_{y_1 y_2}^2 G_{y_2 y_2} \qquad G_{n_2 n_2} = (1 - \gamma_{y_1 y_2}^2) G_{y_2 y_2} \tag{7.19}$$

Hence the output signal-to-noise ratio along path 2 is given by

$$\frac{G_{vv}}{G_{n_2 n_2}} = \frac{\gamma_{y_1 y_2}^2}{1 - \gamma_{y_1 y_2}^2} \tag{7.20}$$

that is, the coherence function between the outputs will determine the output signal-to-noise ratio along path 2 when there is negligible noise in $y_1(t)$.

Next consider the case where $G_{n_1 n_2} = G_{n_2 n_2} = G_{nn} \neq 0$ and

$$H_1 = \exp(-j2\pi f \tau_1) \qquad |H_1| = 1$$
$$H_2 = \exp(-j2\pi f \tau_2) \qquad |H_2| = 1$$

For arbitrary H_1 and H_2, the coherence between the outputs is given by Equation 7.10 as

$$\gamma_{y_1 y_2}^2 = \frac{|H_1|^2 |H_2|^2 G_{xx}^2}{(|H_1|^2 G_{xx} + G_{nn})(|H_2|^2 G_{xx} + G_{nn})} \tag{7.21}$$

For the special case where $|H_1| = |H_2| = 1$,

$$\gamma_{y_1 y_2}^2 = \frac{G_{xx}^2}{(G_{xx} + G_{nn})^2} \tag{7.22}$$

It follows that

$$\frac{G_{xx}}{G_{nn}} = \frac{|\gamma_{y_1 y_2}|}{1 - |\gamma_{y_1 y_2}|} \tag{7.23}$$

where $|\gamma_{y_1 y_2}|$ is the positive square root of $\gamma_{y_1 y_2}^2$. This case corresponds to different time delays τ_1, τ_2 for the two paths but similar extraneous noise.

The quantity $|\gamma_{y_1y_2}|$ now determines the signal-to-noise ratio G_{xx}/G_{nn} which is the same as G_{uu}/G_{nn} or G_{vv}/G_{nn} since $G_{xx} = G_{uu} = G_{vv}$ when $|H_1| = |H_2| = 1$.

7.2   RELATIVE TIME DELAYS AND PROPAGATION DIRECTION

Consider two corrupted noise signals $y_1(t)$ and $y_2(t)$, as pictured in Figure 7.2, where $y_1(t)$ and $y_2(t)$ are given by

$$y_1(t) = x(t) + n_1(t)$$
$$y_2(t) = \alpha x(t - \tau_1) + n_2(t)$$
(7.24)

The constants α and τ_1 represent, respectively, a relative attenuation factor and a time delay. The two extraneous noise terms $n_1(t)$ and $n_2(t)$ are assumed to be uncorrelated with each other and with $x(t)$.

7.2.1   *Estimation of Relative Time Delays*

From Equations 3.15 and 3.30, the cross-correlation and cross-spectral density functions of $y_1(t)$ and $y_2(t)$ as defined in Equation 7.24 are given by

$$R_{y_1y_2}(\tau) = E[y_1(t)y_2(t + \tau)] = \alpha R_{xx}(\tau - \tau_1)$$

$$G_{y_1y_2}(f) = 2 \int_{-\infty}^{\infty} R_{y_1y_2}(\tau)e^{-j2\pi f\tau}\,d\tau = \alpha e^{-j2\pi f\tau_1}G_{xx}(f)$$
(7.25)

while the autocorrelation and autospectra are given by

$$R_{y_1y_1}(\tau) = E[y_1(t)y_1(t + \tau)] = R_{xx}(\tau) + R_{n_1n_1}(\tau)$$
$$G_{y_1y_1}(f) = G_{xx}(f) + G_{n_1n_1}(f)$$
(7.26)

$$R_{y_2y_2}(\tau) = E[y_2(t)y_2(t + \tau)] = \alpha^2 R_{xx}(\tau) + R_{n_2n_2}(\tau)$$
$$G_{y_2y_2}(f) = \alpha^2 G_{xx}(f) + G_{n_2n_2}(f)$$
(7.27)

The formulas for $R_{y_1y_2}(\tau)$ and $G_{y_1y_2}(f)$ in Equation 7.25 are independent of the noise terms $n_1(t)$ and $n_2(t)$. The result for $R_{y_1y_2}(\tau)$ is a function of α, τ_1, and $R_{xx}(\tau)$, and clearly has a maximum value at $\tau = \tau_1$. The time delay τ_1 appears in the cross-spectrum as a linear phase angle given by

$$\theta_{y_1y_2}(f) = 2\pi f\tau_1$$
(7.28)

Hence the time delay τ_1 can be estimated using either the cross-correlation function or the cross-spectral density function defined in Equation 7.25.

The above results are identical in principle to the time delay calculations for nondispersive multiple-path propagation problems developed in Chapter

6, and their applications are limited by many of the same practical considerations. In this case, however, there is no problem with overlapping of adjacent correlation peaks because it is assumed that there is only one path between $y_1(t)$ and $y_2(t)$. On the other hand, it is still desirable here to make the correlation peak at $\tau = \tau_1$ as sharp as feasible to obtain the maximum accuracy in the definition of τ_1. This can often be done by a procedure analogous to that presented in Section 6.1.3 as follows: Let a hypothetical frequency response function $H_{y_1y_2}(f)$ be defined, as in Equation 5.1, by

$$H_{y_1y_2}(f) = \frac{G_{y_1y_2}(f)}{G_{y_1y_1}(f)} \tag{7.29}$$

where $y_1(t)$ is treated as the input and $y_2(t)$ as the output in a single input/single output model. The associated coherence function is

$$\gamma^2_{y_1y_2}(f) = \frac{|G_{y_1y_2}(f)|^2}{G_{y_1y_1}(f)G_{y_2y_2}(f)} \tag{7.30}$$

A *complex coherence function* can now be defined by

$$\gamma_{y_1y_2}(f) = \frac{G_{y_1y_2}(f)}{[G_{y_1y_1}(f)G_{y_2y_2}(f)]^{1/2}} \tag{7.31}$$

where the phase of the function is the phase angle of $G_{y_1y_2}(f)$. From Equations 7.25 and 7.26,

$$H_{y_1y_2}(f) = \frac{\alpha e^{-j2\pi f \tau_1}}{1 + [G_{n_1n_1}(f)/G_{xx}(f)]} \tag{7.32}$$

Thus $H_{y_1y_2}(f)$ is independent of $G_{n_2n_2}(f)$, but is a function of α, τ_1, and the noise-to-signal ratio $G_{n_1n_1}(f)/G_{xx}(f)$. If this ratio is negligible compared to unity, then

$$H_{y_1y_2}(f) \approx \alpha e^{-j2\pi f \tau_1} \tag{7.33}$$

This gives a result independent of $G_{xx}(f)$ and $G_{n_1n_1}(f)$ which has the form of bandwidth-limited white noise. Now the inverse Fourier transform of $H_{y_1y_2}(f)$ can be expected to estimate τ_1 quite closely unless α is small compared to basic instrument and computation errors. Theoretically, if $n_1(t) = 0$, this would give a delta function type result, namely,

$$h_{y_1y_2}(\tau) \approx \alpha\delta(\tau - \tau_1) \tag{7.34}$$

These ideas were illustrated previously in Figure 6.7.

From Equations 7.25 through 7.27, a general complex coherence function between $y_1(t)$ and $y_2(t)$ can be defined by

$$\gamma_{y_1 y_2}(f) = \frac{\alpha e^{-j2\pi f \tau_1}}{[1 + \{G_{n_1 n_1}(f)/G_{xx}(f)\}]^{1/2}[\alpha^2 + \{G_{n_2 n_2}(f)/G_{xx}(f)\}]^{1/2}} \quad (7.35)$$

If the first noise-to-signal ratio $G_{n_1 n_1}(f)/G_{xx}(f)$ is negligible compared to unity, then

$$\gamma_{y_1 y_2}(f) \approx \frac{\alpha e^{-j2\pi f \tau_1}}{[\alpha^2 + \{G_{n_2 n_2}(f)/G_{xx}(f)\}]^{1/2}} \quad (7.36)$$

Hence $\gamma_{y_1 y_2}(f)$ is a function of α, τ_1, and the second noise-to-signal ratio $G_{n_2 n_2}(f)/G_{xx}(f)$. For cases where this second ratio is negligible compared to α^2, Equation 7.36 becomes

$$\gamma_{y_1 y_2}(f) \approx e^{-j2\pi f \tau_1} \quad (7.37)$$

This also has the form of bandwidth-limited white noise and would be preferred over both Equations 7.25 and 7.33 since it is independent of the attenuation factor α as well as $G_{xx}(f)$. However, if $G_{n_2 n_2}(f) \gg G_{xx}(f)$, as occurs in unfavorable situations, then Equations 7.33 and 7.34 would be preferred over Equation 7.36 and its associated inverse Fourier transform.

7.2.2 Determination of Dominant Direction

Suppose the two output signals $y_1(t)$ and $y_2(t)$ in Figure 7.2 represent a pair of orthogonal components as pictured in Figure 7.3. Let $y_c(t)$ be the in-line vector output record with angle ϕ relative to the $y_1(t)$ record, and $y_p(t)$

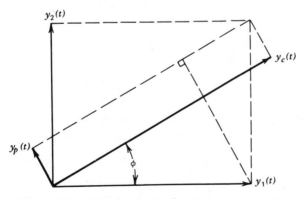

Figure 7.3 Pair of orthogonal output measurements.

be the output record perpendicular to $y_c(t)$ with an angle $\phi + 90$ relative to the $y_1(t)$ record. For any angle ϕ,

$$y_c(t) = y_1(t) \cos \phi + y_2(t) \sin \phi$$
$$y_p(t) = -y_1(t) \sin \phi + y_2(t) \cos \phi \qquad (7.38)$$

Finite Fourier transforms of $y_c(t)$ and $y_p(t)$ over a long record length T yield

$$Y_c(f, T) = Y_1(f, T) \cos \phi + Y_2(f, T) \sin \phi$$
$$Y_p(f, T) = -Y_1(f, T) \sin \phi + Y_2(f, T) \cos \phi \qquad (7.39)$$

The autospectral density functions $G_{cc}(f)$ and $G_{pp}(f)$ can be expressed in terms of $G_{11}(f) = G_{y_1y_1}(f)$, $G_{22}(f) = G_{y_2y_2}(f)$, and $G_{12}(f) = G_{y_1y_2}(f)$ by

$$G_{cc}(f) = G_{11}(f) \cos^2 \phi + G_{22}(f) \sin^2 \phi + 2Re[G_{12}(f)] \sin \phi \cos \phi$$
$$G_{pp}(f) = G_{11}(f) \sin^2 \phi + G_{22}(f) \cos^2 \phi - 2Re[G_{12}(f)] \sin \phi \cos \phi$$
$$(7.40)$$

From Equations 3.33 and 3.40, $G_{11}(f)$, $G_{22}(f)$ and $Re[G_{12}(f)]$ give the variance and cross-variance for any pair of stationary random records $y_1(t)$ and $y_2(t)$ with zero mean values as follows:

$$\sigma_1^2 = E[y_1^2(t)] = \int_0^\infty G_{11}(f) \, df$$

$$\sigma_2^2 = E[y_2^2(f)] = \int_0^\infty G_{22}(f) \, df \qquad (7.41)$$

$$\sigma_{12} = E[y_1(t)y_2(t)] = \int_0^\infty Re[G_{12}(f)] \, df$$

Hence the variances of $y_c(t)$ and $y_p(t)$ are given by

$$\sigma_c^2 = \sigma_1^2 \cos^2 \phi + \sigma_2^2 \sin^2 \phi + 2\sigma_{12} \sin \phi \cos \phi$$
$$\sigma_p^2 = \sigma_1^2 \sin^2 \phi + \sigma_2^2 \cos^2 \phi - 2\sigma_{12} \sin \phi \cos \phi \qquad (7.42)$$

The associated cross-spectra $G_{cp}(f)$ and cross-variance σ_{cp} from Equation 7.39 are

$$G_{cp}(f) = [G_{22}(f) - G_{11}(f)] \sin \phi \cos \phi$$
$$+ G_{12}(f) \cos^2 \phi - G_{21}(f) \sin^2 \phi$$
$$Re[G_{cp}(f)] = [G_{22}(f) - G_{11}(f)] \sin \phi \cos \phi \qquad (7.43)$$
$$+ Re[G_{12}(f)](\cos^2 \phi - \sin^2 \phi)$$
$$\sigma_{cp} = (\sigma_2^2 - \sigma_1^2) \sin \phi \cos \phi + \sigma_{12}(\cos^2 \phi - \sin^2 \phi)$$

Consider now the in-line autospectrum $G_{cc}(f)$ given in Equation 7.40. Observe that $G_{cc}(f)$ is a function of both frequency and direction. The *dominant direction* ϕ_0 will now be defined as that value of ϕ which maximizes $G_{cc}(f)$. To determine ϕ_0, set the partial derivative of $G_{cc}(f)$ with respect to ϕ equal to zero and solve for the resulting value of ϕ. This gives

$$\frac{\partial G_{cc}(f)}{\partial \phi} = 2[G_{22}(f) - G_{11}(f)] \sin \phi \cos \phi$$

$$+ 2Re[G_{12}(f)](\cos^2 \phi - \sin^2 \phi) = 0$$

But

$$\sin 2\phi = 2 \sin \phi \cos \phi \qquad \cos 2\phi = \cos^2 \phi - \sin^2 \phi$$

Hence $\phi = \phi_0(f)$ must satisfy the equation

$$\tan 2\phi_0(f) = \frac{2Re[G_{12}(f)]}{G_{11}(f) - G_{22}(f)} \tag{7.44}$$

This gives two possible answers corresponding to a maximum or a minimum value for $G_{cc}(f)$ so that a check is required. If it is determined that a particular $\phi_0(f)$ gives a minimum value, then the correct dominant direction angle will be $\phi_0(f) \pm 90°$ Note that substitution of $\phi = \phi_0(f)$ into Equation 7.43 results in $Re[G_{cp}(f)] = 0$, hence $\sigma_{cp} = 0$.

A second way to determine a dominant direction is to consider the in-line variance from Equation 7.42, namely

$$\sigma_c^2 = \sigma_1^2 \cos^2 \phi + \sigma_2^2 \sin^2 \phi + 2\sigma_{12} \sin \phi \cos \phi \tag{7.45}$$

where σ_c^2 is independent of frequency and a function only of direction. The dominant direction ϕ_1 can now be defined as that value of ϕ which maximizes σ_c^2. Proceeding as before, the angle ϕ_1 must satisfy the equation

$$\tan 2\phi_1 = \frac{2\sigma_{12}}{\sigma_1^2 - \sigma_2^2} \tag{7.46}$$

This result also must be checked to resolve a possible $90°$ ambiguity. Note that ϕ_1 is independent of frequency and does not provide the detailed frequency characteristics contained in $\phi_0(f)$ from Equation 7.44. A simpler way to obtain this same ϕ_1 is to set σ_{cp} of Equation 7.43 equal to zero.

7.2.3 *Further Relationships*

Let $A(f)$ represent the ratio of the in-line spectrum $G_{cc}(f)$ to the perpendicular spectrum $G_{pp}(f)$, namely

$$A(f) = \frac{G_{cc}(f)}{G_{pp}(f)} \tag{7.47}$$

where $G_{cc}(f)$ and $G_{pp}(f)$ satisfy Equation 7.40. The quantity $A(f)$ is a function of both frequency and direction. A second way to define a dominant direction is by that value of ϕ, denoted by $\phi_2(f)$, which maximizes $A(f)$. By the straightforward operations conducted previously, it is found that $\phi_2(f)$ must satisfy the requirement

$$\tan 2\phi_2(f) = \frac{2Re[G_{12}(f)]}{G_{11}(f) - G_{22}(f)} \tag{7.48}$$

Thus $\phi_2(f)$ is identical with $\phi_0(f)$ of Equation 7.44. The correct dominant direction between the two possible answers will now be given by that value of $\phi_2(f)$ for which $A(f)$ is greater than one.

Referring back to Figures 7.2 and 7.3, for any angle ϕ, an *in-line coherence function* can be defined between the input record $x(t)$ in Figure 7.2 and the in-line vector output record $y_c(t)$ in Figure 7.3. Recalling that $y_c(t)$ is computed from the two output records $y_1(t)$ and $y_2(t)$ in Figure 7.2, it follows that

$$\gamma_{xc}^2(f) = \frac{|G_{xc}(f)|^2}{G_{xx}(f)G_{cc}(f)} \tag{7.49}$$

where $G_{cc}(f)$ satisfies Equation 7.40 and

$$G_{xc}(f) = G_{x1}(f) \cos \phi + G_{x2}(f) \sin \phi \tag{7.50}$$

Thus $\gamma_{xc}^2(f)$ is clearly a function of direction ϕ as well as frequency f, and therefore should be denoted by $\gamma_{xc}^2(f, \phi)$. From the previously determined dominant directions $\phi_0(f)$ or ϕ_1 given in Equations 7.44 and 7.46, one can substitute and obtain

$$\gamma_{xc}^2[f, \phi_0(f)] \quad \text{and} \quad \gamma_{xc}^2(f, \phi_1) \tag{7.51}$$

These coherence functions can then be used to determine the validity of single input/single output linear models between input records $x(t)$ and dominant in-line vector output records $y_c(t)$ in the directions $\phi_0(f)$ or ϕ_1.

7.3 SINGLE SOURCE LOCATION PROBLEMS

A common application of the analysis procedures in Section 7.2 is to the location of an unknown input source emitting energy with a known propagation velocity, based on time delay calculations between pairs of output measurements. Specifically, consider a single source that produces spherically propagating energy which is measured at two output locations with a separation distance d, as illustrated in Figure 7.4a. If the propagation velocity c is

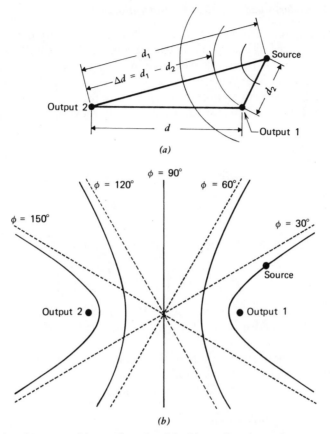

Figure 7.4   Source position and angle of incidence based on phase measurements between two outputs. (*a*) Path length difference. (*b*) Source location.

known, then the time delay τ_1 measured between the outputs $y_1(t)$ and $y_2(t)$ may be interpreted as an angle of incidence ϕ which satisfies

$$\cos \phi = \frac{c\tau_1}{d} \qquad (7.52)$$

However, the source does not lie on the line determined by ϕ. In actuality, the source will lie on a hyperbolic surface given by

$$\Delta d = d_1 - d_2 = c\tau_1 \qquad (7.53)$$

as illustrated in Figure 7.4*b* for any plane passing through the output points. An exact location of the source in three-dimensional space would of course require such measurements from three orthogonal pairs of output locations.

7.3.1 Procedures and Practical Considerations

From Equation 7.25, the time delay τ_1 between two output signals $y_1(t)$ and $y_2(t)$ produced by a single input can be computed using either the cross-correlation or cross-spectral density function of $y_1(t)$ and $y_2(t)$. In practice, the cross-spectrum (coherence and phase) is more commonly used because of certain advantages in identifying output noise problems, to be discussed later. From Equation 7.37, for the ideal case of no extraneous noise in the outputs,

$$\gamma^2_{12}(f) = \gamma^2_{y_1 y_2}(f) = 1 \qquad \theta_{12}(f) = \theta_{y_1 y_2}(f) = 2\pi f \tau_1 \qquad (7.54)$$

where the phase may be written using τ_1 from Equation 7.52 as

$$\theta_{12}(f) = \frac{[2\pi f d \cos \phi]}{c} \qquad (7.55)$$

Hence the angle of incidence of the propagation is given by

$$\phi = \cos^{-1}\left[\frac{c\theta_{12}(f)}{2\pi f d}\right] = \cos^{-1}\left[\frac{c\theta'_{12}(f)}{2\pi d}\right] \qquad (7.56)$$

where $\theta'_{12}(f)$ is the derivative of $\theta_{12}(f)$ with respect to f and has the units of radians/Hz. For nondispersive propagation, $[\theta_{12}(f)/f] = \theta'_{12}(f) = \theta'_{12}$ is a constant.

The application of these results to the location of unmeasured sources is straightforward in theory, but a number of potential problems can limit their effectiveness in practice, including the following:

1. Finite source dimensions.
2. Contaminating output noise.
3. Reverberation and scattering effects.
4. Data acquisition and sampling errors.

These problems are discussed in the next four sections.

7.3.2 Finite Source Dimensions

The results in Equations 7.54 through 7.56 assume that the outputs $y_1(t)$ and $y_2(t)$ are due to an input $x(t)$ representing a single point source. If the source is in fact relatively large compared to its distance from the output locations and is not fully coherent within itself, then the source will actually appear to the outputs as a collection of uncorrelated sources. This situation generally will produce a coherence $\gamma^2_{12}(f)$ that decays with increasing frequency, and a phase $\theta_{12}(f)$ that is somewhat ambiguous. However, except for very nearby sources, the distortion of $\theta_{12}(f)$ will often not be as severe as might be expected.

As an illustration, consider the experiment involving coherence and phase measurements of the acoustic noise from a jet engine, as outlined in Figure 7.5. The measurements were made between two microphones about 40 meters aft and to the side of the engine while operating at full thrust. The microphones were separated by $d = 2.2$ meters and the angle of incidence from the jet noise to the microphone axis was $\phi \approx 23°$. The resulting coherence and phase measurements calculated with a resolution of $B_e = 4$ Hz and $n_d = 256$ averages over the frequency range from near zero to 1200 Hz are shown in Figure 7.6. The source (jet noise) has substantial dimensions at these frequencies (perhaps a meter) and is not highly coherent within itself. Hence the decaying coherence function in Figure 7.6 is expected. Nevertheless, the phase data are relatively clean with an average slope of 0.0374 radians/Hz. From Equation 7.56, this corresponds to $\phi \approx 23°$, in general agreement with the actual geometry of the microphones. Of course, the source distance to microphone separation ratio in this illustration is well over 10. For ratios of less than 10, the phase results might become more confused.

7.3.3 Contaminating Output Noise

From Equation 7.35, the presence of statistically independent (extraneous) noise in the output measurements $y_1(t)$ and $y_2(t)$, typical of instrumentation noise, will reduce the coherence $\gamma_{12}^2(f)$ but will not alter the phase $\theta_{12}(f)$. Contaminating noise in $y_1(t)$ and $y_2(t)$ that is correlated, however, will generally distort the phase data, although meaningful information might still be obtained in some cases.

One of the most common forms of contaminating noise is *diffuse noise* due to numerous independent noise sources that surround the output locations

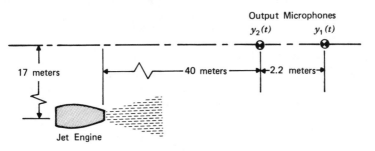

Figure 7.5 Jet noise experiment.

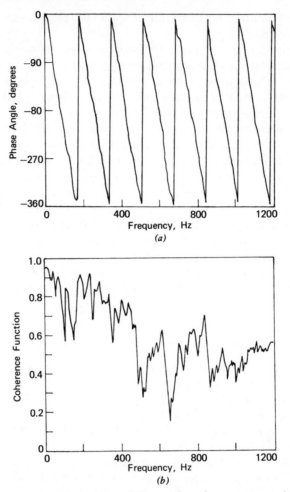

Frequency resolution $B_e = 4$ Hz, number of averages, $n_d = 256$

Figure 7.6 Phase and coherence of jet noise between two outputs. (*a*) Phase function. (*b*) Coherence function. (These data were acquired from experiments performed for the NASA Johnson Space Center under Contract NAS-15231.)

and are picked up in $y_1(t)$ and $y_2(t)$. Such noise can be modeled with reasonable accuracy by an infinite collection of statistically independent noise sources spherically surrounding the output locations, as illustrated in Figure 7.7. The cross-spectrum between $y_1(t)$ and $y_2(t)$ for this case can be shown to be [7.2]

$$G_{12}(f) = G_d(f) \frac{\sin k_0 d}{k_0 d} \tag{7.57}$$

Primary Input

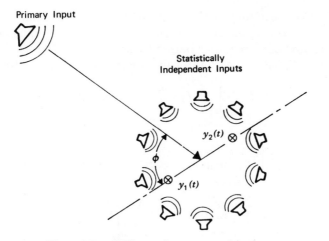

Figure 7.7   Diffuse noise at output locations.

where

$$G_d(f) = \text{diffuse noise autospectra at both } y_1(t) \text{ and } y_2(t)$$

$$k_0 = \left(\frac{2\pi f}{c}\right) = \text{wave number}$$

In terms of coherence and phase, diffuse noise alone appears as

$$\gamma_{12}^2(f) = \left(\frac{\sin k_0 d}{k_0 d}\right)^2 \qquad \theta_{12}(f) = 0 \text{ for } G_{12}(f) \text{ positive} \qquad (7.58)$$
$$= \pi \text{ for } G_{12}(f) \text{ negative}$$

Referring back to Equations 7.25 and 7.52, and assuming an attenuation factor of $\alpha = 1$, the cross-spectrum between $y_1(t)$ and $y_2(t)$ due to a single point source can be written as

$$G_{12}(f) = G_p(f)[\cos (k_t d) - j \sin (k_t d)] \qquad (7.59)$$

where

$$G_p(f) = \text{point source autospectra at both } y_1(t) \text{ and } y_2(t)$$

$$k_t = \left(\frac{2\pi f}{c}\right) \cos \phi = \text{trace wave number}$$

The cross-spectrum of the diffuse noise combined with the single point source is given by the sum of Equations 7.57 and 7.59, specifically,

$$G_{12}(f) = G_p(f) \cos (k_t d) + G_d(f) \frac{(\sin k_0 d)}{k_0 d} - j G_p(f) \sin (k_t d) \quad (7.60)$$

where $G_p(f)$ is the autospectrum of the point source contribution and $G_d(f)$ is the autospectrum of the diffuse noise contribution at the output locations. The total autospectrum at each output location will be $G_p(f) + G_d(f)$. The coherence and phase between this pair of outputs will then be given by

$$\gamma_{12}^2(f) = [1 + R(f)]^{-2}\left[\left\{R(f)\frac{(\sin k_0 d)}{k_0 d} + \cos(k_t d)\right\}^2 + \sin^2(k_t d)\right]$$

(7.61a)

$$\theta_{12}(f) = \tan^{-1}\left[\frac{\sin(k_t d)}{\{R(f)(\sin k_0 d)/k_0 d\} + \cos(k_t d)}\right]$$

(7.61b)

where

$$R(f) = \left[\frac{G_d(f)}{G_p(f)}\right]$$

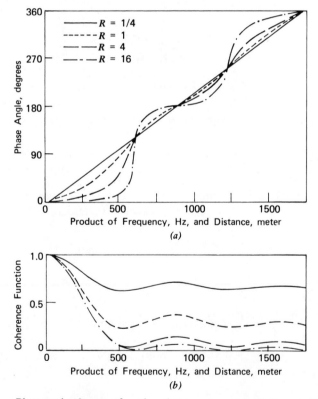

Figure 7.8 Phase and coherence functions between two outputs for combined diffuse and point source inputs. (a) Phase functions. (b) Coherence functions.

Plots of Equation 7.61 for various values of R and $\phi = 45°$ are shown in Figure 7.8. Note in these plots that the phase between the receivers reaches $\pm n\pi$ radians, $n = 1, 2, 3, \ldots$, at a common frequency independent of the ratio R. From Equation 7.61b, $\sin(k_t d) = 0$ corresponds to a phase shift of 0, π, 2π, and so on. In other words, $180°$ phase shifts occur when

$$k_t d = n\pi = \left(\frac{2\pi f}{c}\right) d \cos \phi \qquad n = 1, 2, 3, \ldots \qquad (7.62)$$

Hence at those frequencies f_n satisfying Equation 7.62, corresponding to phase shifts of $180°$, $360°$, and so on, the incident angle of the incoming waves from the point source is given by

$$\phi = \cos^{-1}\left[\frac{nc}{2f_n d}\right] \qquad (7.63)$$

independent of the contaminating diffuse noise, exactly as stated in Equation 7.56 with $\theta_{12}(f) = \pi$ and $n = 1$.

To verify the above results, consider the wind tunnel acoustic noise experiment [7.3] illustrated in Figure 7.9. Two microphones with streamlining to suppress wind noise were mounted in the wind tunnel test section with a separation distance of $d = 0.356$ meter. The angle of incidence for acoustic waves propagating down the tunnel was $\phi = 45.7°$. The phase and coherence measured between the two microphones during tunnel operation with a flow velocity of $c = 80$ meters/sec is shown in Figure 7.10. These measurements were made using a resolution of $B_e = 5$ Hz and $n_d = 256$ averages. Note the similarity of the measured data to the theoretical results in Figure 7.8. The data clearly indicate the presence of waves propagating from a downstream source plus heavy diffuse noise contamination at the output microphone locations. The diffuse noise was due to the distributed flow noise in the test section as well as reverberation effects, to be discussed later. Now,

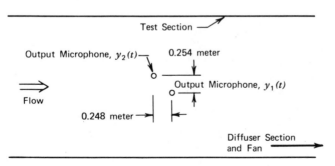

Figure 7.9   Wind tunnel acoustic experiment.

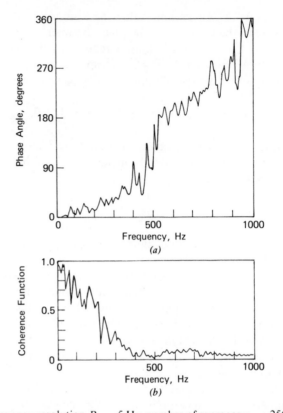

Frequency resolution $B_e = 5$ Hz, number of averages $n_d = 256$

Figure 7.10 Phase and coherence functions between microphones in wind tunnel test section with flow velocity of 80 meters/sec. (*a*) Phase function. (*b*) Coherence function. (These data were acquired from experiments performed for the NASA Ames Research Center under Contract NAS2-8382.)

the first $180°$ phase shift occurs at $f = 520$ Hz. The propagation velocity from a downstream source in this case should be $c = c_0 - c_1$, where $c_0 = 340$ meters/sec (the speed of sound in air at room temperature) and $c_1 = 80$ meters/sec. Hence $c = 260$ meters/sec, and from Equation 7.63, the angle of incidence is given by

$$\phi = \cos^{-1}\left[\frac{260}{2(520)(0.356)}\right] = 45.4°$$

which agrees very well with the actual $\phi = 45.7°$ in Figure 7.9.

7.3.4 *Reverberation and Scattering Effects*

Reverberation refers to that situation where the output locations are surrounded by reflecting surfaces causing both $y_1(t)$ and $y_2(t)$ to be heavily contaminated by repeated reflections. It can be viewed in the context of a propagation medium which is a bounded system that supports standing waves; that is, a system with numerous lightly damped normal modes, as defined in Section 1.3.4. Reverberation obviously will distort the phase between two output measurements $y_1(t)$ and $y_2(t)$ since they will see propagating waves from directions other than the line-of-sight direction to the input. Not so obvious is the fact that the coherence measurement $\gamma_{12}^2(f)$ may also be distorted under certain conditions. Specifically, in the context of the propagation medium being a lightly damped resonant system, as the number of normal modes in the analysis resolution bandwidth B_e becomes large and the damping ζ becomes small, it can be shown [7.4] that the coherence function between two output measurements $y_1(t)$ and $y_2(t)$ will approach

$$\gamma_{12}^2(f) = \left(\frac{\sin 2\pi f d/c}{2\pi f d/c} \right)^2 \tag{7.64}$$

which is exactly the same as the coherence function for diffuse noise defined in Equation 7.58.

A plausible solution to this problem is to suppress reverberation to the maximum possible extent. Actually, this is not as difficult as one might first expect since it is the reflections aft of the output measurements (downstream from the input) that do the most damage. Hence the placement of absorption material aft of the output location will often suppress the reverberation problem to an acceptable degree. This is demonstrated by the simple laboratory acoustic experiment shown in Figure 7.11 where an input acoustic source and two output microphones are completely enclosed to produce a reverberant field. By simply adding 0.07 meter of acoustic absorption material aft of the microphones, the reverberation effects are suppressed to the point where the measured coherence and phase between the two microphone

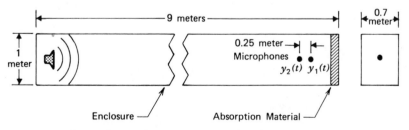

Figure 7.11 Reverberant acoustic noise experiment.

outputs are similar, on the average, to the results expected under ideal conditions, as illustrated by the plots in Figure 7.12.

The above noted procedure of appropriate employment of absorption does pose a hazard if the absorption is incorrectly placed at a location that is not aft of the outputs. Specifically, it can be shown [7.5] that the phase between two points in a semi-reverberant environment tends to indicate the direction of power flow out of the reverberant region, independent of the source location. Absorption in a restricted location causes an outflow of power at that location, hence it can erroneously influence the phase data if the location of the absorption is, say, to the side of the outputs rather than aft of them. Further discussions of the reverberation problem are presented in Section 9.2.3.

Beyond the reverberation problem, difficulties often arise in the practical use of the procedures outlined here when obstructions or discontinuities lie in the path between the input and output locations. Such obstructions or

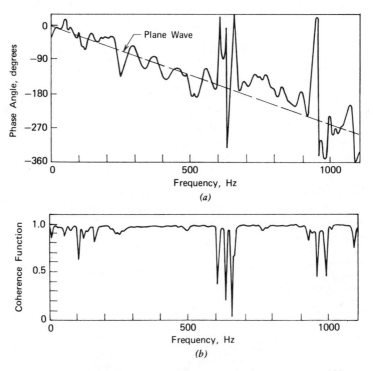

Frequency resolution $B_e = 2.5$ Hz, number of averages $n_d = 128$

Figure 7.12 Phase and coherence functions between two acoustic outputs in a reverberant field. (*a*) Phase function. (*b*) Coherence function.

discontinuities may cause scattering of the propagating energy, as discussed previously in Section 6.1.4. Scattering is particularly detrimental to accurate phase measurements and must be avoided if fully meaningful results are to be obtained.

7.3.5 Data Acquisition and Sampling Errors

Proper calibration of data acquisition and analysis equipment is always important in any signal processing problem. For the source location problem outlined in Section 7.3, however, a proper phase calibration of the instrumentation is particularly critical since the computed phase data are the primary information of interest. Phase errors can arise anywhere throughout a data acquisition/analysis system, including the transducers, signal conditioning equipment, and data recording/reproduction operations. Of course, the absolute phase shift in an individual measurement is of no concern; it is the relative phase shift between the two instrumentation channels producing $y_1(t)$ and $y_2(t)$ that is detrimental.

Relative phase errors occur in two forms: (a) static and (b) dynamic. Static (bias) phase errors can be easily identified by end-to-end calibration of the measurement instrumentation and then removed from the analyzed results by simple subtraction. Repeating the measurements with the transducers reversed in position can also be used to obtain corrected results. Specifically, from Equation 7.28, if there is a static phase error between $y_1(t)$ and $y_2(t)$ corresponding to a time delay $\tau_e = [\theta_e(f)/2\pi f]$, then the correct phase will be given by

$$\theta_{12}(f)_c = \frac{\theta_{12}(f) + \theta_{21}(f)}{2} = \frac{2\pi f(\tau_1 + \tau_e) + 2\pi f(\tau_1 - \tau_e)}{2} = 2\pi f \tau_1 \quad (7.65)$$

Dynamic phase errors are more difficult to identify and correct. The simplest approach here is to repeat the end-to-end phase calibration several times to obtain a measure of the phase stability.

From Chapter 11, the random error in the phase estimates determined by Equation 7.28 due to statistical sampling considerations is given in terms of the standard deviation of the estimate $\hat{\theta}_{12}(f)$, by

$$\sigma[\hat{\theta}_{12}(f)] \approx \sin^{-1}\left\{\frac{[1 - \gamma_{12}^2(f)]^{1/2}}{|\gamma_{12}(f)|\sqrt{2n_d}}\right\} \quad (7.66)$$

where $\sigma[\hat{\theta}_{12}(f)]$ is measured in radians and n_d is the number of averages used in the spectral computations. For the special case where the term in the { } brackets is small, say, less than 0.1, Equation 7.66 simplifies to

$$\sigma[\hat{\theta}_{12}(f)] \approx \frac{[1 - \gamma_{12}^2(f)]^{1/2}}{|\gamma_{12}(f)|\sqrt{2n_d}} \quad (7.67)$$

In practice, the unknown coherence function on the right side of the equation would be replaced with the estimated coherence function calculated from the data.

It is clear from Equation 7.67 that the calculation of highly accurate phase estimates demands either a very strong value of the coherence between $y_1(t)$ and $y_2(t)$ or a large number of averages in the computations. For example, assume that one wishes to define the phase angle to within a standard deviation of $\pm 1°$ or 0.0174 radians, as might be required to firmly locate a source. If the measured coherence between the two signals is, say $\hat{\gamma}_{12}^2 = 0.9$, then the necessary number of averages would be

$$n_d \approx \frac{[1 - 0.9]}{2(0.9)(0.0174)^2} \approx 184$$

This requires a substantial but not unreasonable amount of data. On the other hand, if the measured coherence were only $\hat{\gamma}_{12}^2 = 0.50$, then

$$n_d \approx \frac{[1 - 0.5]}{2(0.5)(0.0174)^2} \approx 1652$$

which might constitute an unreasonable record length requirement.

In reality, the phase estimation error problem is not quite as severe as suggested above because a single analysis will usually produce a collection of phase estimates at many different frequencies, for example, as shown in Figure 7.6. The errors in phase estimates at different frequencies are statistically independent. Hence, assuming nondispersive propagation, if a straight line is fit to several estimates at different frequencies, the error in the fitted line will be less than the error in any individual phase estimate. When curve fitting procedures are used, however, one must be very careful not to include phase estimates at frequencies where they are prone to bias errors due to contaminating noise, as discussed in Section 7.3.3, or where reverberation and/or scattering effects are present, as covered in Sections 7.3.4 and 9.2.3.

7.3.6   Periodic Sources

The source location procedures in Section 7.3.1 are particularly effective when the source is periodic (assuming its period is known) because the desired results occur at discrete frequencies. Hence narrow-band analysis procedures can be used to suppress the effects of contaminating noise. To illustrate this fact, consider the experiment involving pressure measurements on the fuselage of a reciprocating engine, propeller-driven aircraft, as outlined in Figure 7.13. The impinging pressures on the fuselage were measured with an array of flush-mounted microphones, as shown in Figure 7.13a. A typical pressure time history record during static operation is presented in

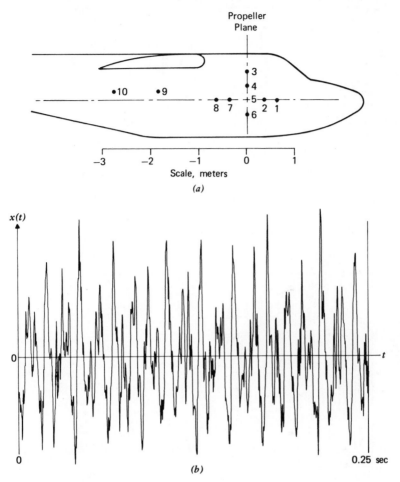

Figure 7.13 Propeller airplane fuselage pressure experiments. (*a*) Transducer locations. (*b*) Typical pressure time history record.

Figure 7.13*b*. Note that the pressure record is quite complex since it represents the net sum of pressures due to the propeller, exhaust, and other miscellaneous sources. Typical phase and coherence data, calculated between locations 4 and 5 in the plane of the propeller with a resolution of $B_e = 4\,\text{Hz}$ and $n_d = 126$ averages are shown in Figure 7.14. These results also appear to be quite complicated at first glance.

If attention is now restricted to the pressures due to the propeller only, the results in Figure 7.14 can be greatly simplified since it is known that

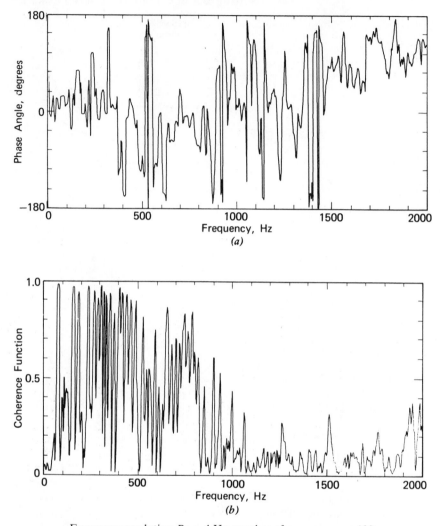

Frequency resolution $B_e = 4$ Hz, number of averages $n_d = 100$

Figure 7.14 Phase and coherence functions between pressures at locations 4 and 5 on propeller airplane during static operation at 2600 rpm. (*a*) Phase function. (*b*) Coherence function. (The data in Figures 7.14 through 7.16 were acquired from experiments performed for the NASA Langley Research Center under Contract NAS1-14611.)

the primary propeller contributions occur at the propeller-blade passage frequency and all harmonics thereof, given by

$$f_k = k \text{ (engine rpm) (propeller gear ratio) (number of blades)}$$
$$k = 1, 2, 3, \ldots .$$

(7.68)

The phase information in Figure 7.14a at the blade passage frequencies only is shown for two different engine rpms in Figure 7.15a. Note that the data now reveal a linear phase shift characteristic of nondispersive propagation. Also note that the slope of the phase data is different when the engine rpm is changed. Furthermore, the propagation velocities from Equation 7.55 correspond to the tip speed of the propeller blade projected to the transducer locations. This result is most informative since it suggests that the propeller-induced pressures at these locations are due to aerodynamic forces rather than acoustic noise which must always propagate at the speed of sound.

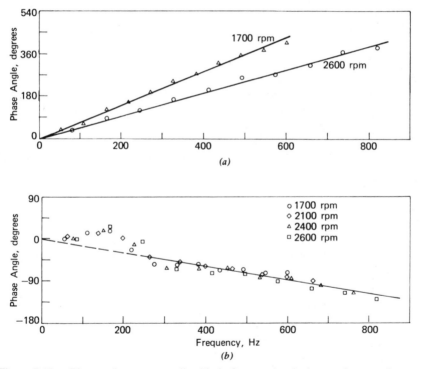

Figure 7.15 Phase values at propeller blade frequencies during static operation at different engine speeds. (a) Between locations 4 and 5 in plane of propeller. (b) Between locations 1 and 2 along longitudinal axis.

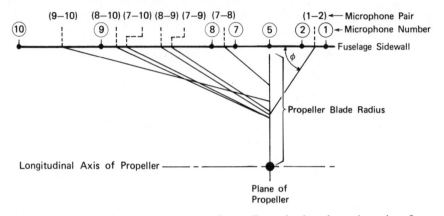

Figure 7.16   Effective source location of propeller noise based on phase data from various pairs of transducers.

To carry the illustration a step further, the phase data calculated at the propeller-blade passage frequencies between locations 1 and 2 along the longitudinal axis forward of the propeller are shown for four different engine speeds in Figure 7.15*b*. For this case, the data above 200 Hz all display linear phase shifts with a common slope independent of the engine rpm, suggesting that the pressures are due to acoustic noise (the pressures below 200 Hz are still dominated by aerodynamic forces). From Equation 7.56, for acoustic propagation at $c = 340$ meters/sec and noting that $d = 0.254$ meter and $\theta_{12}(f) = \pi/2$ at $f = 610$ Hz, the angle of incidence of the acoustic wave must be

$$\phi = \cos^{-1}\left[\frac{340(\pi/2)}{2\pi(610)(0.254)}\right] = 57°$$

The propeller, of course, constitutes a distributed noise source, but calculations of this type can be used to identify an equivalent acoustic center for the propeller noise using the procedures detailed in Section 7.3.1. Such calculations from the phase data for various microphone pairs in Figure 7.13 are shown in Figure 7.16. Note that the data from all locations point to an equivalent acoustic center in the same general region on the propeller.

7.4   SYSTEM IDENTIFICATION FROM OUTPUT MEASUREMENTS

Chapters 4 and 5 discuss the identification of system properties based on the analysis of input/output data. Situations commonly arise, however, where input excitations cannot be accurately measured or conveniently simulated. This is often true of large civil engineering structures such as tall buildings, long bridges, and offshore platforms. If an artificial excitation can be applied, then the dynamic characteristics of such structures can be calculated by procedures of the type described in Section 5.1.2. In many cases, however, reasonable estimates of fundamental dynamic properties can be deduced by appropriate analysis of only output measurements representing the response of the structure to a natural dynamic environment that is random in character, for example, the wind loads on buildings, the traffic excitation on bridges, and the wave forces on offshore structures [7.6, 7.7].

7.4.1   *Identification of Resonance Frequencies*

From Equation 4.8, when a lightly damped structure is subjected to a random excitation, the output autospectrum at any response point will reach a maximum at frequencies where either the excitation spectrum peaks or the frequency response function of the structure peaks. From Section 1.3, narrow-band peaks in the frequency response function of lightly damped mechanical systems occur at those frequencies corresponding to system normal modes (resonance frequencies). Hence peaks in response spectra generally can be assumed to represent either peaks in the excitation spectrum or normal modes of the structure.

To distinguish between output spectral peaks which are due to structural vibration modes, as opposed to peaks in the input spectrum, one can take advantage of the fact that all points on a structure responding in a lightly damped normal mode of vibration will be either in phase or 180° out of phase with one another, depending only on the shape of the normal mode as described in Section 1.3.4. This is illustrated for the bending modes of a cantilever-beam-type structure in Figure 7.17. The two longitudinally spaced points shown in this example move in phase for lateral vibration in the first and third bending modes, and 180° out of phase in the second and forth bending modes. A distinction between bending and torsional modes can be assisted by additional phase measurements between two laterally spaced transducers on opposite sides of the structure at the same longitudinal location. In the lateral plane, the output signals from any two such transducers must be in phase for all bending modes and 180° out of phase for all torsional modes, assuming the modes are completely decoupled. At those frequencies where a peak in the output spectrum is due to a spectral peak in the excitation, and not a resonant response of the structure, the phase data between two output measurements will usually be something other than zero or 180°.

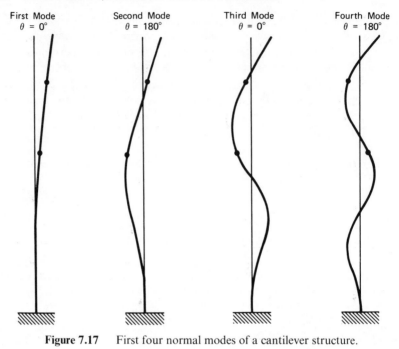

First Mode
$\theta = 0°$

Second Mode
$\theta = 180°$

Third Mode
$\theta = 0°$

Fourth Mode
$\theta = 180°$

Figure 7.17   First four normal modes of a cantilever structure.

To illustrate these principles, consider an experiment involving a large offshore structure, shown schematically in Figure 7.18. The structure, which extends below the water line and is firmly attached to the sea bottom, was instrumented with accelerometers producing output measurements of the structural motion in a common direction as follows: $y_1(t)$ on the right inboard leg near the top of the structure, $y_2(t)$ on the right inboard leg near the

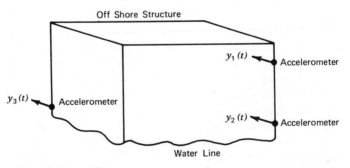

Off Shore Structure

$y_1(t)$ — Accelerometer

$y_3(t)$ — Accelerometer

$y_2(t)$ — Accelerometer

Water Line

Figure 7.18   Offshore structure wave-force response experiment.

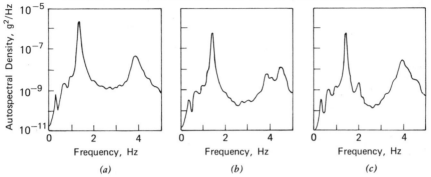

Frequency, Hz   Frequency, Hz   Frequency, Hz

(a)   (b)   (c)

Frequency resolution $B_e = 0.04$ Hz, number of averages $n_d = 55$

Figure 7.19   Autospectra of offshore structure response. (a) $y_1(t)$. (b) $y_2(t)$. (c) $y_3(t)$.

water line, and $y_3(t)$ on the left outboard leg near the water line. All data were analyzed using a resolution bandwidth of $B_e = 0.04$ Hz and $n_d = 55$ averages.

The autospectra of the signals from the three accelerometers in Figure 7.18 due to normal wave-force loading on the structure are shown over the frequency range from near zero to 5 Hz in Figure 7.19. These data reveal several spectral peaks at all three locations, but by themselves do not provide sufficient information to identify desired normal modes of the structure. The cross-spectrum magnitude and phase, along with coherence calculations, are shown in Figure 7.20. These results are more significant. Specifically, the magnitudes of the cross-spectra between $y_1(t)$ and $y_2(t)$, as well as between $y_2(t)$ and $y_3(t)$, both reveal three dominant peaks at about 1.4, 3.9, and 4.5 Hz. There are also some smaller peaks at frequencies below 1 Hz, but these probably represent spectral peaks in the wave-force excitation. For the three dominant peaks, the phase of the cross-spectra is either zero or 180°, as expected for a resonant response of the structure. The phase data are summarized in Table 7.1. The results at 1.4 Hz indicate a bending (sway) mode of the structure since all measurements are in phase. At 3.9 Hz, the response

Table 7.1

Phase Results for Response of Offshore Structure

Measurement Pair	Transducer Separation	Relative Phase at Dominant Frequencies		
		1.4 Hz	3.9 Hz	4.5 Hz
$y_1(t)$ vs. $y_2(t)$	Longitudinal	0	0	180
$y_2(t)$ vs. $y_3(t)$	Lateral	0	180	0

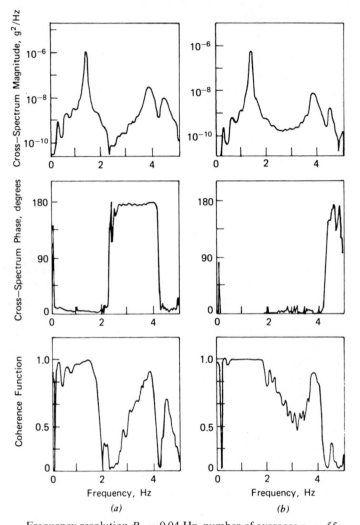

Frequency resolution $B_e = 0.04$ Hz, number of averages $n_d = 55$

Figure 7.20 Cross-spectra and coherence functions between outputs on offshore structure. (a) $y_1(t)$ versus $y_2(t)$. (b) $y_2(t)$ versus $y_3(t)$.

is in phase over the longitudinal separation and 180° out of phase for the lateral separation, which confirms a torsional mode. The results at 4.5 Hz are the opposite; that is, measurements are out of phase for the longitudinal separation and in phase for the lateral separation, indicating a higher order bending mode at this frequency.

The coherence functions for $y_1(t)$ versus $y_2(t)$ and for $y_2(t)$ versus $y_3(t)$ are also shown in Figure 7.20. These coherence functions are needed to establish the random errors in the phase estimates, which in turn establish the statistical significance of a discrepancy between estimated and anticipated phase values as given by Equation 7.66. However, the coherence functions can also be helpful in determining normal mode shapes, as discussed in the next section. Note that the coherence functions in Figure 7.20 tend to peak at the identified normal mode frequencies. This results from the fact that the normal modes appear as narrow-band peaks in the output spectra, and thus the signal-to-noise ratio in the calculations is maximized at these frequencies, as detailed in Section 7.1.

7.4.2 *Determination of Normal Mode Shapes*

Having identified normal mode frequencies, the order and shape of each normal mode can theoretically be determined if a sufficient number of transducers are spread over the structure of interest. From the measurements in any given direction, the relative shape of the ith normal mode would be approximated at the r discrete points where measurements are made by

$$\phi_i(y_j) = [G_{y_j y_j}(f_i)]^{1/2} \qquad \begin{array}{l} i = 1, 2, 3, \ldots \\ j = 1, 2, \ldots, r \end{array} \qquad (7.69)$$

Here, $G_{y_j y_j}(f_i)$ is the output autospectral density value at the ith normal mode frequency and the jth location. In general, the minimum number of measurements r required to firmly identify the ith order normal mode is $r = i$, but a much larger number of measurements may be needed to accurately define the ith normal mode shape in detail. Of course, if some other information can be brought to bear, for example, normal mode shape predictions from analytical studies of the structure, then a fewer number of measurements might be used in conjunction with analytical results to identify specific modes and estimate their shapes.

Even when a sufficient number of measurements can be made, there are several limitations on the application of Equation 7.69. The spectral peaks in output autospectra measurements actually occur at resonance frequencies and not the undamped natural frequencies desired in normal mode determinations. From Equation 1.56, these frequencies are in close agreement only for very small values of the modal damping ratio ζ. Hence application of the technique should be limited to cases where the damping ratio is small, say $\zeta < 0.05$.

Significant extraneous noise in the output measurements and coupling among the normal modes will distort these results. However, problems of this type will be revealed by coherence and phase measurements among the output data. From Equation 7.10, extraneous noise in the measurement at a specific location will cause the coherence function between that measurement and all other measurements to be less than unity. Furthermore, a coupling between normal modes will cause the phase data among at least some of the measurements to be other than zero or 180°. As a general rule, the autospectrum at a specific location should not be used to define a normal mode shape unless the measurement produces near unity coherence and near zero or 180° phase with all other output measurements. For example, referring to Figures 7.19 and 7.20, the autospectral density values at 1.4 Hz should provide accurate indications of the relative modal deflections at the three measurement points involved here because the coherence values among the measurements at this frequency are near unity and the phases are near zero. At 3.9 Hz, the phase data are still good, but the coherence values are about $\gamma^2 = 0.9$. This suggests that the autospectral density values are slightly contaminated by extraneous noise, but would still provide reasonably accurate estimates of the relative modal deflections. At 4.5 Hz, however, the coherence values fall to as low as $\gamma^2 = 0.25$, meaning one or more of the autospectra estimates at this frequency includes substantial extraneous noise and would severely overestimate the relative modal deflection.

7.4.3   *Estimation of Structural Damping*

For each normal mode identified by the procedures outlined in Section 7.4.1, an equivalent viscous damping ratio for the mode can sometimes be estimated from output auto and/or cross-spectra data. Specifically, four conditions must exist: (a) The autospectrum of the excitation $G_{xx}(f)$ must be reasonably uniform over the frequency of the normal mode, (b) the modal damping ζ_i must be small, (c) the analysis resolution bandwidth B_e must be much smaller than the half-power point bandwidth B_i of the mode, and (d) the mode must not overlap heavily with neighboring modes. As rules of thumb,

(a) $G_{xx}(f) \approx$ Constant for $f_i - 3B_i \leq f \leq f_i + 3B_i$
(b) $\zeta_i < 0.05$
(c) $B_e < 0.2B_i$
(d) $(f_i - f_{i-1}) > 2(B_i + B_{i-1})$

If these conditions are met, then an equivalent viscous damping ratio for the *ith* mode can be estimated from Equation 1.59 by

$$\zeta_i = \frac{B_i}{2f_i} \tag{7.70}$$

where B_i is the half-power point bandwidth of the spectral peak associated with the *ith* mode. It is preferable to calculate B_i from the cross-spectrum between two output measurements rather than from autospectrum data because the cross-spectrum is less contaminated by noise.

To illustrate the application of Equation 7.70, assume that the data in Figure 7.19 meet requirements (*a*) and (*b*) stated above. The half-power point bandwidths and estimated damping ratios of the three normal modes indicated by the dominant peaks in the cross-spectra from Figure 7.20 are

$$f_1 = 1.4 \text{ Hz} \qquad B_1 = 0.10 \text{ Hz} \qquad \zeta_1 = 0.036$$
$$f_2 = 3.9 \text{ Hz} \qquad B_2 = 0.30 \text{ Hz} \qquad \zeta_2 = 0.038$$
$$f_3 = 4.5 \text{ Hz} \qquad B_3 = 0.23 \text{ Hz} \qquad \zeta_3 = 0.025$$

However, the first spectral peak at $f_1 = 1.4$ Hz does not meet the resolution requirement since $B_e = 0.04$ Hz $= 0.4B_1$, which is larger than the maximum desired value of $0.2B_1$, and the spectral peaks at $f_2 = 3.9$ Hz and $f_3 = 4.5$ Hz do not meet the overlap requirement since $f_3 - f_2 = 0.6$ Hz $= 1.1(B_3 + B_2)$, which is smaller than the minimum desired value of $2(B_3 + B_2)$. Both of these deficiencies cause damping to be overestimated, hence the values of ζ computed above are probably too large.

REFERENCES

7.1 Carter, G. C., "Time Delay Estimation," NUSC TR-5335, *Naval Underwater Systems Center*, New London, Connecticut, April 1976.

7.2 Cron, B. J., and Sherman, C. H., "Spatial-Correlation Functions for Various Noise Models," *Journal of the Acoustical Society of America*, Vol. 34, No. 11, p. 1732, November 1962.

7.3 Piersol, A. G., "Use of Coherence and Phase Data Between Two Receivers in Evaluation of Noise Environments," *Journal of Sound and Vibration*, Vol. 56, No. 2, p. 215, 1978.

7.4 Morrow, C. T., "Point-to-Point Correlation of Sound Pressures in Reverberant Chambers," *Shock and Vibration Bulletin*, No. 39, 1969.

7.5 Blake, W. K., and Waterhouse, R. V., "The Use of Cross-Spectral Density Measurements in Partially Reverberant Sound Fields," *Journal of Sound and Vibration*, Vol. 54, No. 4, p. 589, 1977.

7.6 Abel-Ghaffer, A. M., and Housner, G. W., "Ambient Vibration Tests of Suspension Bridge," *Journal of the Engineering Mechanics Division*, ASCE, Vol. 104, No. EM5, p. 983, 1978.

7.7 Begg, R. D., Mackenzie, A. C., Dodds, C. J., and Loland, O., "Structural Integrity Monitoring Using Digital Processing of Vibration Signals," *Proceedings. Offshore Technology Conference*, OTC 2549, Vol. 2, 1976.

CHAPTER 8

MULTIPLE INPUT/OUTPUT RELATIONSHIPS

This chapter is concerned with fundamental input/output relationships for multiple input problems. It is assumed here that inputs are records from stationary (ergodic) random or transient random processes with zero mean values, and that systems are constant parameter linear systems.

8.1 MULTIPLE INPUT/OUTPUT SYSTEMS

Consider q clearly defined measurable inputs $x_i(t)$, $i = 1, 2, \ldots, q$, which pass through q constant parameter linear systems with frequency response functions $H_i(f)$, $i = 1, 2, \ldots, q$, so as to produce a single measured output $y(t)$. The output $y(t)$ will be the sum of the ideal predicted linear outputs $v_i(t)$, $i = 1$, $2, \ldots, q$, plus all possible deviations from the ideal model included in the unknown $n(t)$, as shown in Figure 8.1. Optimum choices of $H_i(f)$ in the least-squares sense will make $n(t)$ uncorrelated (uncoherent) with $x_i(t)$. This situation is a realistic physical case where it is desired to estimate $H_i(f)$ and $n(t)$ from measurements of $x_i(t)$ and $y(t)$.

If the same inputs produce a number of different outputs of interest, as in multiple input/multiple output problems, then a similar diagram should be set up for each output point. In this way, the multiple input/multiple output problem can be treated as a combination of separate and simpler multiple input/single output problems.

To emphasize the importance of conducting this analysis in the frequency domain, all derivations will be carried out here without requiring the computation of autocorrelation or cross-correlation functions. Such time domain expressions can be derived if desired, but they are not needed or deemed useful for many engineering applications that depend on determining precisely how systems behave as a function of frequency. They could be important, however, for other engineering applications where different time delays via different transmission paths have to be resolved, as discussed in Chapter 6.

For convenience and clarity, two-sided spectral quantities $S(f)$ will be used in derivations here rather than the one-sided quantities $G(f)$, as defined in Equations 3.45 to 3.47. All of the resulting formulas will be valid when the $S(f)$ terms are replaced by the corresponding $G(f)$. Furthermore, to simplify notation, all finite Fourier transforms will be denoted by capital letters without notation for their frequency and record-length dependence. For example, the governing finite Fourier transform equations for the output $y(t)$ in Figure 8.1 are

$$Y(f, T) = \sum_{i=1}^{q} V_i(f, T) + N(f, T) \tag{8.1}$$

$$V_i(f, T) = H_i(f)X_i(f, T) \qquad i = 1, 2, \ldots, q \tag{8.2}$$

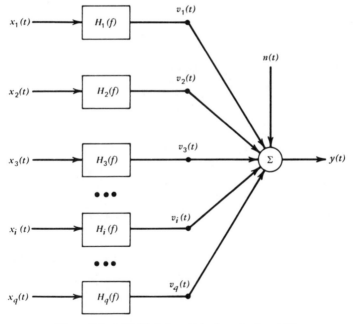

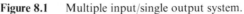

Figure 8.1   Multiple input/single output system.

where $Y(f, T)$, $V_i(f, T)$, $N(f, T)$, and $X_i(f, T)$ are the Fourier transforms of $y(t)$, $v_i(t)$, $n(t)$, and $x_i(t)$, respectively, computed over long sample records of length T. These equations without the frequency and record length notation are

$$Y = \sum_{i=1}^{q} V_i + N \qquad V_i = H_i X_i \qquad i = 1, 2, \ldots, q \qquad (8.3)$$

8.1.1   Basic Input/Output Relationships

Starting with Equation 8.3 and multiplying Y by Y^*, it follows that

$$Y^*Y = \sum_{i=1}^{q} \sum_{j=1}^{q} H_i^* H_j X_i^* X_j + N^*N \qquad (8.4)$$

plus two other terms that average to zero when $n(t)$ and $x_i(t)$ are uncorrelated. Also,

$$Y^*Y = \sum_{i=1}^{q} H_i^* X_i^* Y + N^*Y \qquad (8.5)$$

From the definition in Equation 3.45, taking expected values of Equation 8.4 and dividing by T in the limit as T approaches infinity yields the auto-spectrum of $y(t)$ as

$$S_{yy} = \sum_{i=1}^{q} \sum_{j=1}^{q} H_i^* H_j S_{ij} + S_{nn} \qquad (8.6)$$

Using Equation 8.5, this is equivalent to

$$S_{yy} = \sum_{i=1}^{q} H_i^* S_{iy} + S_{ny} \qquad (8.7)$$

Spectral quantities can be obtained here using a finite T as follows:

$$S_{yy} = \frac{E[Y^*Y]}{T}, \qquad S_{ij} = \frac{E[X_i^* X_j]}{T}$$

$$S_{nn} = \frac{E[N^*N]}{T}, \qquad S_{iy} = \frac{E[X_i^* Y]}{T} \qquad (8.8)$$

$$S_{ny} = \frac{E[N^*Y]}{T}, \qquad S_{in} = \frac{E[X_i^* N]}{T}$$

Note that the quantity

$$S_{ny} = S_{nn} \qquad \text{when } S_{in} = 0 \qquad (8.9)$$

Also, since $S_{yy}^* = S_{yy}$ and $S_{iy}^* = S_{yi}$, it follows that

$$S_{yy} = \sum_{i=1}^{q} H_i S_{yi} + S_{ny} \qquad (8.10)$$

Multiplication of Equation 8.3 by X_i^* after first changing the index of summation to j shows

$$X_i^* Y = \sum_{j=1}^{q} H_j X_i^* X_j + X_i^* N \qquad (8.11)$$

Expected values and divisions by T then give

$$S_{iy} = \sum_{j=1}^{q} H_j S_{ij} \qquad i = 1, 2, \ldots, q \qquad (8.12)$$

when $S_{in} = 0$. Equation 8.12 is a set of q equations that can be solved for the q unknown $\{H_j\}$ when the model is well defined and when all of the S_{iy} and S_{ij} terms are known. Recommended computational procedures are discussed in Chapter 10.

8.1.2   Optimum Frequency Response Functions

From Equation 8.3, the noise term N and its complex conjugate can be written as

$$N = Y - \sum_{i=1}^{q} H_i X_i \qquad (8.13)$$

Then

$$N^* = Y^* - \sum_{j=1}^{q} H_j^* X_j^* \qquad (8.14)$$

The product of Equations 8.13 and 8.14 gives

$$N^* N = Y^* Y - \sum_{i=1}^{q} H_i Y^* X_i - \sum_{j=1}^{q} H_j^* X_j^* Y + \sum_{i=1}^{q} \sum_{j=1}^{q} H_j^* H_i X_j^* X_i \quad (8.15)$$

Taking expected values and dividing by T, it follows that

$$S_{nn} = S_{yy} - \sum_{i=1}^{q} H_i S_{yi} - \sum_{j=1}^{q} H_j^* S_{jy} + \sum_{i=1}^{q} \sum_{j=1}^{q} H_j^* H_i S_{ji} \qquad (8.16)$$

This gives the form of S_{nn} for *any* choice of constant parameter linear systems described by H_j. The *optimum systems* will now be defined as those systems which minimize S_{nn} over all possible choices of H_j. As shown previously in Section 5.1.1, the optimum H_j can be obtained by setting

$$\frac{\partial S_{nn}}{\partial H_j} = 0 \quad \text{and} \quad \frac{\partial S_{nn}}{\partial H_j^*} = 0 \tag{8.17}$$

where in the first case, H_j^* is held fixed, and in the second case, H_j is held fixed. By this technique, one finds here that

$$\frac{\partial S_{nn}}{\partial H_j^*} = -S_{jy} + \sum_{i=1}^{q} H_i S_{ji} = 0 \tag{8.18}$$

Thus

$$S_{jy} = \sum_{i=1}^{q} H_i S_{ji} \quad j = 1, 2, \ldots, q \tag{8.19}$$

which is exactly the same form as Equation 8.12. It follows now from Equation 8.13 that $S_{jn} = 0$ for all j.

If it is assumed that the input records $x_i(t)$ are mutually uncorrelated, then Equations 8.12 and 8.19 reduce to

$$S_{jy} = H_j S_{jj} \quad j = 1, 2, \ldots, q \tag{8.20}$$

This shows that the overall system of Figure 8.1 is now merely a combination of simple single input/single output systems, where the frequency response function H_j of the jth path can be determined without regard to the other inputs that may be present. Of course these other inputs affect the coherence function and the statistical accuracy in estimating H_j from measured data.

8.1.3 Two Input/Single Output Systems

The special case of a two input/single output system will now be considered in some detail to illustrate matters of concern in the general case of q inputs. The two inputs $x_1(t)$ and $x_2(t)$ in Figure 8.2 are arbitrary stationary random or transient random records that may be correlated. From measurements of $x_1(t)$, $x_2(t)$, and $y(t)$, it is desired to determine $H_1(f)$, $H_2(f)$, and $S_{nn}(f)$.

For this two input case, Equation 8.12 states that

$$\begin{aligned} S_{1y} &= H_1 S_{11} + H_2 S_{12} \\ S_{2y} &= H_1 S_{21} + H_2 S_{22} \end{aligned} \tag{8.21}$$

Now assume that the coherence between $x_1(t)$ and $x_2(t)$ is

$$\gamma_{12}^2 = \frac{|S_{12}|^2}{S_{11} S_{22}} \neq 1 \tag{8.22}$$

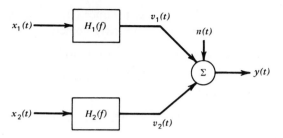

Figure 8.2   Two input/single output system.

In other words, $x_1(t)$ and $x_2(t)$ are not fully correlated. Then, general solutions for H_1 and H_2 are

$$H_1 = \frac{S_{22}S_{1y} - S_{12}S_{2y}}{S_{11}S_{22} - |S_{12}|^2} = \frac{S_{1y}\left[1 - \dfrac{S_{12}S_{2y}}{S_{22}S_{1y}}\right]}{S_{11}[1 - \gamma_{12}^2]}$$

$$H_2 = \frac{S_{11}S_{2y} - S_{21}S_{1y}}{S_{11}S_{22} - |S_{12}|^2} = \frac{S_{2y}\left[1 - \dfrac{S_{21}S_{1y}}{S_{11}S_{2y}}\right]}{S_{22}[1 - \gamma_{12}^2]}$$

(8.23)

Observe that if $S_{12} = 0$ and $\gamma_{12}^2 = 0$, then special solutions are

$$H_1 = \frac{S_{1y}}{S_{11}} \quad \text{and} \quad H_2 = \frac{S_{2y}}{S_{22}}$$

(8.24)

representing the usual single input/output results developed in Chapter 4.

The situation of fully correlated inputs where $\gamma_{12}^2 = 1$ must be handled separately. A coherence function of unity between $x_1(t)$ and $x_2(t)$ implies complete linear dependence. Hence one would consider a linear system existing between them, as illustrated in Figure 8.3. The first input $x_1(t)$ can

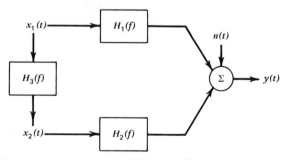

Figure 8.3   Two input problem with fully correlated inputs.

take two different transmission paths to produce $y(t)$, and a single frequency response function $H(f)$ will relate $y(t)$ to $x_1(t)$, namely

$$H(f) = H_1(f) + H_2(f)H_3(f) \qquad (8.25)$$

Assume now that γ_{12}^2 does not equal zero or unity but satisfies $0 < \gamma_{12}^2 < 1$. From Equation 8.6, S_{yy} is given by

$$S_{yy} = |H_1|^2 S_{11} + H_1^* H_2 S_{12} + H_2^* H_1 S_{21} + |H_2|^2 S_{22} + S_{nn} \qquad (8.26)$$

The terms $H_1^* H_2 S_{12} + H_2^* H_1 S_{21}$ are due to the interaction occurring between $x_1(t)$ and $x_2(t)$, and would not be present if $S_{12} = 0$. The ordinary coherence functions between each input and the output are

$$\gamma_{1y}^2 = \frac{|S_{1y}|^2}{S_{11} S_{yy}} = \frac{|H_1 S_{11} + H_2 S_{12}|^2}{S_{11} S_{yy}}$$

$$\gamma_{2y}^2 = \frac{|S_{2y}|^2}{S_{22} S_{yy}} = \frac{|H_1 S_{21} + H_2 S_{22}|^2}{S_{22} S_{yy}} \qquad (8.27)$$

These equations are difficult to interpret, and it is not obvious that they must be bounded between zero and one.

Consider the coherent output spectrum between $x_1(t)$ and $y(t)$. This is given by

$$\gamma_{1y}^2 S_{yy} = \frac{|H_1 S_{11} + H_2 S_{12}|^2}{S_{11}}$$

$$= \frac{|H_1|^2 S_{11}^2 + H_1 H_2^* S_{11} S_{21} + H_1^* H_2 S_{11} S_{12} + |H_2|^2 |S_{12}|^2}{S_{11}}$$

$$= |H_1|^2 S_{11} + H_1 H_2^* S_{21} + H_1^* H_2 S_{12} + \gamma_{12}^2 |H_2|^2 S_{22} \qquad (8.28)$$

The product $\gamma_{1y}^2 S_{yy}$ represents the spectral output at $y(t)$ due to $x_1(t)$ passing through *both* H_1 and H_2. Similarly, by interchanging the indices 1 and 2,

$$\gamma_{2y}^2 S_{yy} = |H_2|^2 S_{22} + H_2 H_1^* S_{12} + H_2^* H_1 S_{21} + \gamma_{12}^2 |H_1|^2 S_{11} \qquad (8.29)$$

represents the spectral output at $y(t)$ due to $x_2(t)$ passing through *both* H_1 and H_2.

Finally, for the special case when $S_{12} = 0$, Equations 8.26 and 8.27 become

$$S_{yy} = |H_1|^2 S_{11} + |H_2|^2 S_{22} + S_{nn} \qquad (8.30)$$

$$\gamma_{1y}^2 = \frac{|H_1|^2 S_{11}}{S_{yy}} \qquad \gamma_{2y}^2 = \frac{|H_2|^2 S_{22}}{S_{yy}} \qquad (8.31)$$

Here, $\gamma_{1y}^2 S_{yy}$ represents the spectral output at $y(t)$ due to $x_1(t)$ passing only through H_1 since the other path does not occur for $x_1(t)$. A similar interpretation exists for $\gamma_{2y}^2 S_{yy}$ with respect to $x_2(t)$ and H_2. When $S_{nn} = 0$, it follows that

$$\gamma_{1y}^2 + \gamma_{2y}^2 = 1 \tag{8.32}$$

Now if $x_1(t)$ and $x_2(t)$ are of equal importance in producing $y(t)$, then $\gamma_{1y}^2 = \gamma_{2y}^2$ and

$$\gamma_{1y}^2 = \gamma_{2y}^2 = 0.5 \tag{8.33}$$

Thus the ordinary coherence functions will be less than unity even though there are perfect linear systems between each input and the output. This result occurs because the effect of the second input is to act like extraneous noise at the output.

8.2 MULTIPLE COHERENCE FUNCTIONS

The multiple coherence function is a direct extension of the concept of ordinary coherence which provides a measure of the linear dependence between a collection of q inputs and an output, independent of the correlation among the inputs.

8.2.1 Definition and Basic Relationships

Referring to Figure 8.1, the *multiple coherence function* between the output $y(t)$ and all of the inputs $x_i(t)$, $i = 1, 2, \ldots, q$, that produce $y(t)$ is defined as the ratio of the ideal predicted linear output spectrum S_{vv} divided by the total measured output spectrum S_{yy}. Thus if

$$S_{yy} = S_{vv} + S_{nn} \tag{8.34}$$

where S_{nn} is the extraneous noise output spectrum, it follows that

$$\gamma_{y:x}^2 = \frac{S_{vv}}{S_{yy}} = \frac{S_{yy} - S_{nn}}{S_{yy}} = 1 - \frac{S_{nn}}{S_{yy}} \tag{8.3}$$

The notation $y:x$ is that portion of $y(t)$ which is due to $x_1(t), x_2(t), \ldots, x_q($ From Equations 8.6 and 8.7, the general form of S_{vv} is

$$S_{vv} = \sum_{i=1}^{q} \sum_{j=1}^{q} H_i^* H_j S_{ij} = \sum_{i=1}^{q} H_i^* S_{iy} \tag{8.}$$

Since $S_{vv} \le S_{yy}$, $S_{nn} \le S_{yy}$ and all terms are non-negative, it follows for frequencies f that

$$0 \le \gamma_{y:x}^2 \le 1 \tag{8.}$$

From Equation 8.35, the product of the multiple coherence function and the output spectrum yields

$$\gamma_{y:x}^2 S_{yy} = S_{vv} = S_{yy} - S_{nn} \tag{8.38}$$

This product, called the *multiple coherent output (power) spectrum*, is a direct extension of the ordinary coherent output (power) spectrum of Equation 4.30. It represents the fractional portion of the output spectrum S_{yy} due to all of the measured inputs $x_i(t)$, $i = 1, 2, \ldots, q$, that produce $y(t)$. The *output noise spectrum* S_{nn} not due to any of the $x_i(t)$ is given here by

$$S_{nn} = (1 - \gamma_{y:x}^2)S_{yy} \tag{8.39}$$

Application of these results requires that input noise be negligible compared to input signals, and that all of the input and output records be measured using a common time base so as to preserve their proper relative phase relationships.

8.2.2 Special Cases

To further clarify the interpretations of multiple coherence functions and their correspondence with ordinary coherence functions, consider the following special cases. First, for a single input,

$$S_{vv} = |H_1|^2 S_{11} = H_1^* S_{1y} = \gamma_{1y}^2 S_{yy}$$

and

$$S_{nn} = (1 - \gamma_{1y}^2)S_{yy}$$

It then follows from Equation 8.35 that

$$\gamma_{y:x}^2 = \gamma_{1y}^2 \tag{8.40}$$

Thus the ordinary coherence function is a special case of the multiple coherence function.

For the two input case,

$$\begin{aligned} S_{vv} &= H_1^* S_{1y} + H_2^* S_{2y} \\ &= H_1^*(H_1 S_{11} + H_2 S_{12}) + H_2^*(H_1 S_{21} + H_2 S_{22}) \end{aligned}$$

The multiple coherence function for the general two input case is then given by

$$\gamma_{y:x}^2 = \frac{H_1^*(H_1 S_{11} + H_2 S_{12}) + H_2^*(H_1 S_{21} + H_2 S_{22})}{S_{yy}} \tag{8.41}$$

where $S_{yy} = S_{vv} + S_{nn}$. This equation is difficult to interpret when left in this form. However, for this special case where $S_{12} = 0$, Equation 8.41 becomes easier to interpret. Specifically, from Equation 8.31,

$$\gamma_{y:x}^2 = \frac{|H_1|^2 S_{11} + |H_2|^2 S_{22}}{S_{yy}} = \gamma_{1y}^2 + \gamma_{2y}^2 \tag{8.42}$$

In other words, Equation 8.42 states that when the two inputs are uncorrelated, the multiple coherence function is the sum of the ordinary coherence functions between each input and the output. This result extends to any number of inputs when the inputs are mutually uncorrelated with each other, as will now be proved.

For the general case of q mutually uncorrelated inputs, $S_{ij} = 0$ for $i \neq j$ and Equation 8.36 reduces to

$$S_{vv} = \sum_{i=1}^{q} |H_i|^2 S_{ii}$$

Now

$$\gamma_{y:x}^2 = \frac{S_{vv}}{S_{yy}} = \frac{\sum_{i=1}^{q} |H_i|^2 S_{ii}}{S_{yy}} \tag{8.43}$$

But

$$\gamma_{iy}^2 = \frac{|H_i|^2 S_{ii}}{S_{yy}} \qquad i = 1, 2, \ldots, q \tag{8.44}$$

Hence

$$\gamma_{y:x}^2 = \sum_{i=1}^{q} \gamma_{iy}^2 \tag{8.45}$$

It should be emphasized again that this result applies only to situations where the inputs are mutually uncorrelated. A broader formula involving partial coherence functions is required when the inputs are not mutually uncorrelated, as derived in Equation 10.69 of Chapter 10.

8.3   CONDITIONED SPECTRAL ANALYSIS

In the following introduction to more advanced multiple input/output relationships, it is helpful to view the analysis in terms of conditioned records where the equivalent of least-squares prediction operations are used to remove the linear effects of one or more input records from the problem. For clarity, the basic ideas are developed here for the special case of a two input/one output system. These ideas are generalized and extended to arbitrary multiple input systems with appropriate computational algorithms in Chapter 10.

8.3.1 *Definitions and Basic Relationships*

Consider the two input model in Figure 8.2 where $x_1(t)$ and $x_2(t)$ are correlated, but not perfectly correlated, such that the coherence between them at all frequencies is $0 < \gamma_{12}^2 < 1$. Of interest now is the linear dependence between $x_1(t)$ and $y(t)$ when all correlated effects of $x_2(t)$ are removed from the problem, or conversely between $x_2(t)$ and $y(t)$ with the correlated effects of $x_1(t)$ removed. In physical terms, this is equivalent to determining the dependence between one of the inputs and the output when the other input is turned off, *assuming all correlated effects between the two inputs originated from the input that is turned off.* We will first develop the case where the correlated effects of $x_1(t)$ are removed from the analysis. The same results apply when $x_2(t)$ is removed by interchanging $x_2(t)$ for $x_1(t)$.

Let $X_{2 \cdot 1} = X_{2 \cdot 1}(f, T)$ denote the finite Fourier transform over a long record length T of the input $x_2(t)$ with the linear effects of $x_1(t)$ removed from $x_2(t)$. Let $Y_{y \cdot 1} = Y_{y \cdot 1}(f, T)$ denote a similar finite Fourier transform of $y(t)$ with the linear effects of $x_1(t)$ removed from $y(t)$. These two quantities can be viewed as noise terms in the two special single input/single output models shown in Figure 8.4. The frequency response function $L_{12} = L_{12}(f)$ in Figure 8.4 is the optimum linear system to predict $x_2(t)$ from $x_1(t)$, and $L_{1y} = L_{1y}(f)$ is the optimum linear system to predict $y(t)$ from $x_1(t)$. Note that subscripts in L_{12} and L_{1y} place the input index *before* the output index to agree with the physical fact that the input should precede the output.

The solutions for L_{12} and L_{1y} are given by Equation 5.15 as

$$L_{12} = \frac{S_{12}}{S_{11}} \tag{8.46}$$

$$L_{1y} = \frac{S_{1y}}{S_{11}} \tag{8.47}$$

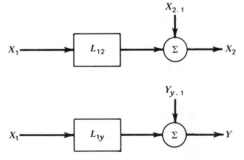

Figure 8.4   Special single input/single output models for two input system.

From Figure 8.4 it is clear that

$$X_{2 \cdot 1} = X_2 - L_{12} X_1 = X_2 - \left(\frac{S_{12}}{S_{11}}\right) X_1 \tag{8.48}$$

$$Y_{y \cdot 1} = Y - L_{1y} X_1 = Y - \left(\frac{S_{1y}}{S_{11}}\right) X_1 \tag{8.49}$$

These two quantities represent the input and output of the conditioned input/output model of Figure 8.5, where L_{2y} is the optimum linear system to predict $Y_{y \cdot 1}$ from $X_{2 \cdot 1}$. The quantity $N = Y_{y \cdot 1, 2}$ is the finite Fourier transform of $y(t)$ conditioned on both $x_1(t)$ and $x_2(t)$. This is the same as the finite Fourier transform of $n(t)$ in Figure 8.2. The system L_{2y} is the same as the system H_2 in Figure 8.2 for any value of γ_{12}^2 because $X_{2 \cdot 1}$ can pass only through H_2 to produce $Y_{y \cdot 1}$; the path through H_1 is removed by the data processing when linear effects of $x_1(t)$ are subtracted from $x_2(t)$.

8.3.2 Conditioned Spectral Density Functions

To simplify further derivations and in agreement with Equation 8.8, define

$$S_{22 \cdot 1} = \frac{E[X_{2 \cdot 1}^* X_{2 \cdot 1}]}{T}$$

$$S_{yy \cdot 1} = \frac{E[Y_{y \cdot 1}^* Y_{y \cdot 1}]}{T} \tag{8.50}$$

$$S_{2y \cdot 1} = \frac{E[X_{2 \cdot 1}^* Y_{y \cdot 1}]}{T}$$

The quantities $S_{22 \cdot 1}$ and $S_{yy \cdot 1}$ are the *conditioned (residual) autospectra* of $x_2(t)$ and $y(t)$, respectively, when the linear effects of $x_1(t)$ are removed from $x_2(t)$ and $y(t)$. The quantity $S_{2y \cdot 1}$ is the *conditioned (residual) cross-spectrum* between $x_2(t)$ and $y(t)$ when the linear effects of $x_1(t)$ are removed from *both* $x_2(t)$ and $y(t)$.

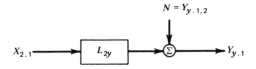

Figure 8.5 Conditioned input/output model for two input system.

From single input/single output considerations, the solution for L_{2y} in Figure 8.5 is

$$L_{2y} = \frac{S_{2y \cdot 1}}{S_{22 \cdot 1}} \tag{8.51}$$

This is a shorthand way to write the result for H_2 given in Equation 8.23, as will be proved shortly. The noise term in Figure 8.5 is given by

$$Y_{y \cdot 1, 2} = Y_{y \cdot 1} - L_{2y} X_{2 \cdot 1} = Y_{y \cdot 1} - \left(\frac{S_{2y \cdot 1}}{S_{22 \cdot 1}}\right) X_{2 \cdot 1} \tag{8.52}$$

and by definition,

$$S_{yy \cdot 1, 2} = \frac{E[Y^*_{y \cdot 1, 2} Y_{y \cdot 1, 2}]}{T} \tag{8.53}$$

In other words, $S_{yy \cdot 1, 2}$ is the conditioned (residual) autospectrum of $y(t)$ when the linear effects of *both* $x_1(t)$ and $x_2(t)$ are removed from $y(t)$ by optimum least-squares techniques. Relationships between Equations 8.8, 8.50, and 8.53 will now be derived.

Substitution of Equations 8.48 and 8.49 into Equation 8.50 and use of Equation 8.8 show that

$$
\begin{aligned}
S_{yy \cdot 1} &= \frac{E\left[\left(Y^* - \dfrac{S_{y1}}{S_{11}} X_1^*\right)\left(Y - \dfrac{S_{1y}}{S_{11}} X_1\right)\right]}{T} \\[2mm]
&= S_{yy} - \left(\frac{S_{1y}}{S_{11}}\right) S_{y1} - \left(\frac{S_{y1}}{S_{11}}\right) S_{1y} + \frac{|S_{1y}|^2}{S_{11}^2} S_{11} \\[2mm]
&= S_{yy} - \frac{|S_{1y}|^2}{S_{11}} = S_{yy}(1 - \gamma_{1y}^2)
\end{aligned}
\tag{8.54}
$$

where

$$\gamma_{1y}^2 = \frac{|S_{1y}|^2}{S_{11} S_{yy}} \tag{8.55}$$

The result in Equation 8.54 is exactly the same as the output noise in single input/single output problems, where $x_1(t)$ is the input and $y(t)$ is the output. A special case of Equation 8.54 when $y(t) = x_2(t)$ gives

$$S_{22 \cdot 1} = S_{22}(1 - \gamma_{12}^2) \tag{8.56}$$

where

$$\gamma_{12}^2 = \frac{|S_{12}|^2}{S_{11}S_{22}} \tag{8.57}$$

In the same way, it follows that

$$S_{2y\cdot 1} = \frac{E\left[\left(X_2^* - \frac{S_{21}}{S_{11}}X_1^*\right)\left(Y - \frac{S_{1y}}{S_{11}}X_1\right)\right]}{T}$$

$$= S_{2y} - \left(\frac{S_{1y}}{S_{11}}\right)S_{21} - \left(\frac{S_{21}}{S_{11}}\right)S_{1y} + \frac{(S_{21})(S_{1y})}{S_{11}^2}S_{11}$$

$$= S_{2y} - \left(\frac{S_{1y}}{S_{11}}\right)S_{21} \tag{8.58}$$

A special case of Equation 8.58 when $x_2(t) = y(t)$ shows

$$S_{yy\cdot 1} = S_{yy} - \left(\frac{S_{1y}}{S_{11}}\right)S_{y1} = S_{yy}(1 - \gamma_{1y}^2) \tag{8.59}$$

in agreement with Equation 8.54. Substitution of Equations 8.58 and 8.56 into Equation 8.51 gives

$$L_{2y} = \frac{S_{2y}\left[1 - \dfrac{S_{21}S_{1y}}{S_{11}S_{2y}}\right]}{S_{22}[1 - \gamma_{12}^2]} \tag{8.60}$$

exactly the same as Equation 8.23 for H_2.

It should be emphasized that expected value operations on conditioned Fourier transforms in Equation 8.50 are not needed to compute these terms. Instead, one should merely use the algebraic operations on the ordinary spectral terms in Equations 8.54, 8.56, and 8.58. However, simpler alternate ways to define the conditioned spectra of Equations 8.50 and 8.53 are given in terms of conditioned Fourier transforms as follows:

$$(T)S_{2y\cdot 1} = E[X_2^* Y_{y\cdot 1}] = E[X_{2\cdot 1}^* Y] \tag{8.61}$$

$$(T)S_{yy\cdot 1, 2} = E[Y^* Y_{y\cdot 1, 2}] = E[Y_{y\cdot 1, 2}^* Y] \tag{8.62}$$

Special cases of $S_{2y\cdot 1}$ give $S_{22\cdot 1}$ and $S_{yy\cdot 1}$. Equation 8.61 shows that it is sufficient to remove the linear effects of $x_1(t)$ from either $x_2(t)$ or $y(t)$ alone, leaving the other record unchanged, to obtain $S_{2y\cdot 1}$. Equation 8.62 shows that it is sufficient to remove the linear effects of $x_1(t)$ and $x_2(t)$ from $y(t)$ in only one of the terms to be averaged to obtain $S_{yy\cdot 1, 2}$.

Proofs of these results, using Equations 8.48 and 8.49 in Equation 8.61, are as follows:

$$S_{2y \cdot 1} = \frac{E\left[X_2^*\left(Y - \frac{S_{1y}}{S_{11}} X_1\right)\right]}{T}$$

$$= S_{2y} - \left(\frac{S_{1y}}{S_{11}}\right) S_{21}$$

$$S_{2y \cdot 1} = \frac{E\left[\left(X_2^* - \frac{S_{21}}{S_{11}} X_1^*\right) Y\right]}{T}$$

$$= S_{2y} - \left(\frac{S_{21}}{S_{11}}\right) S_{1y}$$

These results are exactly the same as Equation 8.58. Similarly, using Equation 8.52 in Equation 8.62 proves

$$S_{yy \cdot 1, 2} = \frac{E\left[Y^*\left(Y_{y \cdot 1} - \frac{S_{2y \cdot 1}}{S_{22 \cdot 1}} X_{2 \cdot 1}\right)\right]}{T}$$

$$= S_{yy \cdot 1} - \left(\frac{S_{2y \cdot 1}}{S_{22 \cdot 1}}\right) S_{y2 \cdot 1}$$

$$S_{yy \cdot 1, 2} = \frac{E\left[\left(Y_{y \cdot 1}^* - \frac{S_{y2 \cdot 1}}{S_{22 \cdot 1}} X_{2 \cdot 1}^*\right) Y\right]}{T}$$

$$= S_{yy \cdot 1} - \left(\frac{S_{y2 \cdot 1}}{S_{22 \cdot 1}}\right) S_{2y \cdot 1}$$

These results are exactly the same as follows from Equation 8.53.

8.4 PARTIAL COHERENCE FUNCTIONS

From the conditioned spectral density quantities defined in Section 8.3, special types of coherence functions can now be calculated that help establish the cause of a linear dependence indicated by an ordinary coherence function. Such coherence functions for conditioned data are called *partial coherence functions*. The definitions and interpretations of partial coherence functions will first be developed for two input/single output systems and will then be extended to the general case of multiple inputs.

8.4.1   Two Input/Single Output Systems

Referring back to the two input/single output system in Figure 8.2, the partial coherence function $\gamma^2_{2y\cdot1}$ between $x_2(t)$ and $y(t)$, when the linear effects of $x_1(t)$ are removed from both $x_2(t)$ and $y(t)$, is now defined as the ordinary coherence function between the conditioned (residual) random variables $X_{2\cdot1}$ and $Y_{y\cdot1}$. Specifically, from Equation 8.50

$$\gamma^2_{2y\cdot1} = \frac{|S_{2y\cdot1}|^2}{S_{22\cdot1}S_{yy\cdot1}} \tag{8.63}$$

The partial coherence function indicates as a function of frequency whether there is a linear relationship between $X_{2\cdot1}$ and $Y_{y\cdot1}$ in Figure 8.5.

From Equations 8.54, 8.56, and 8.58, a general expression for the partial coherence function of Equation 8.63 is

$$\gamma^2_{2y\cdot1} = \frac{|S_{2y}S_{11} - S_{1y}S_{21}|^2}{S^2_{11}S_{22}S_{yy}(1 - \gamma^2_{12})(1 - \gamma^2_{1y})} \tag{8.64}$$

For the special case of uncorrelated inputs where $S_{12} = 0$ and $\gamma^2_{12} = 0$, this reduces to

$$\gamma^2_{2y\cdot1} = \frac{|S_{2y}|^2}{S_{22}S_{yy}(1 - \gamma^2_{1y})} = \frac{\gamma^2_{2y}}{1 - \gamma^2_{1y}} \tag{8.65}$$

assuming $\gamma^2_{1y} \neq 1$. It follows that

$$(1 - \gamma^2_{1y})(1 - \gamma^2_{2y\cdot1}) = 1 - \gamma^2_{1y} - \gamma^2_{2y} \tag{8.66}$$

For a practical engineering application of these matters, consider the fluid dynamic turbulence problem in a torque converter, as schematically illustrated in Figure 8.6. Hydraulic fluid is moved by the pump through stator blades and then drives a turbine. The turbulence velocities were measured using hot wire anemometers at three locations; $x_1(t)$ is forward of the pump inboard, $x_2(t)$ is forward of the turbine inboard, and $y(t)$ is forward of the turbine outboard. All analyses were performed over the frequency range from zero to 3000 Hz using a resolution bandwidth of $B_e = 50$ Hz and $n_d = 64$ averages.

The autospectra of the three turbulence velocity measurements with the mean velocity rendered zero are shown in Figure 8.7. Note that the spectrum of $x_1(t)$ forward of the pump displays a sharp peak at about 400 Hz. This corresponds to the blade passage frequency of the pump. Similarly, the spectra of $x_2(t)$ and $y(t)$ forward of the turbine have sharp peaks at about 300 Hz and harmonics thereof that represent turbine blade passage frequencies. Further note that all three spectra have a small peak at 1920 Hz. The

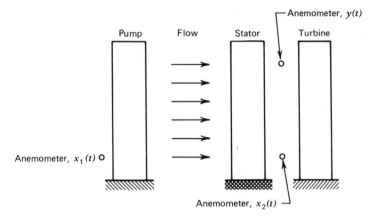

Figure 8.6   Measurement locations in a torque converter.

data were time scaled by $32:1$ during the processing so that 1920 Hz corresponded to 60 Hz in real frequency, that is, the peak at 1920 Hz represents power line pickup.

Of interest in this problem is the linear dependence of the turbulence into the turbine over the span of the blades, that is, between $x_2(t)$ and $y(t)$. As a first step in evaluating this dependence, the ordinary coherence function between $x_2(t)$ and $y(t)$ was calculated, with the result shown in Figure 8.8a. Sharp peaks appear in $\hat{\gamma}_{2y}^2(f)$ at the turbine blade passage tones, as would be expected. There is also a very sharp peak at the transposed power-line frequency of 1920 Hz, again as would be expected; nothing is better correlated than two sine waves at identical frequencies. The only reason that $\hat{\gamma}_{2y}^2(f)$ does not reach unity at 1920 Hz is because of random noise leakage through the 50 Hz wide resolution filter. Beyond these expected coherence peaks due to tones in the data, there is a broad-band coherence that grows stronger with increasing frequency. The question now is whether this broad-band coherence is due to local disturbances generated by the turbine, or a general disturbance (acoustic noise) that propagates in the fluid through the torque converter.

Since the measurement $x_1(t)$ made forward of the pump does not see local disturbances at the turbine, the above question is easily answered by conditioning out of $x_2(t)$ and $y(t)$ all correlated effects with $x_1(t)$. The resulting partial coherence function is shown in Figure 8.8b. Note that there are still significant peaks in the partial coherence at the turbine blade passage frequency and its harmonics since these are indeed genuine local disturbances. However, the high-frequency coherence has largely disappeared, confirming

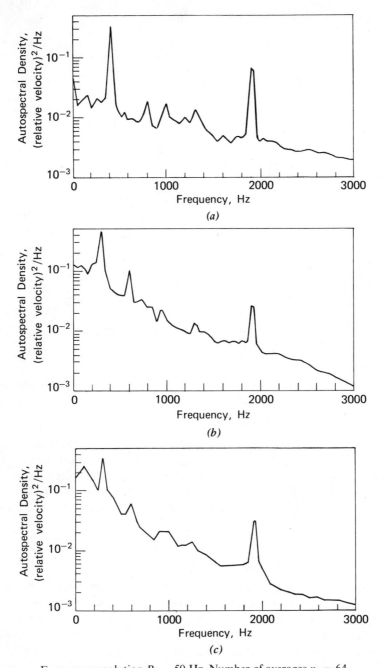

Frequency resolution $B_e = 50$ Hz, Number of averages $n_d = 64$

Figure 8.7 Autospectra of turbulence in torque converter experiment. (*a*) Pump inboard. (*b*) Turbine inboard. (*c*) Turbine outboard.

205

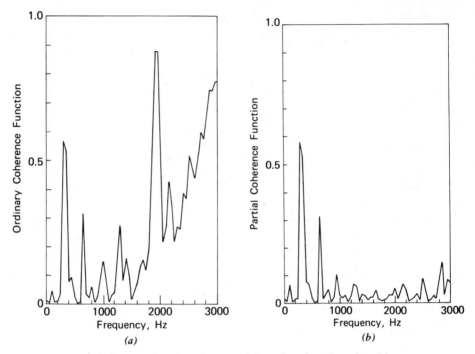

Figure 8.8 Coherence functions between inboard and outboard turbine measurements. (*a*) Ordinary coherence function. (*b*) Partial coherence function.

that the indicated linear dependence between $x_2(t)$ and $y(t)$ at these frequencies represented only their joint response to a common disturbance propagating through the torque converter. Also note that the coherence peak at 1920 Hz has disappeared. Since the power-line pickup occurred in all three measurement channels, the partial coherence calculation simply conditioned it out of the results.

8.4.2 Multiple Input/Single Output Systems

Consider the original model of Figure 8.1 with q arbitrary inputs $x_i(t)$, $i = 1, 2, \ldots, q$, and the output $y(t)$ where all of these $(q + 1)$ records are measured. Consider next any subset of the input records, where the linear effects of r number of the given input records are removed from this subset by optimum linear least-squares techniques. Assume that the linear effects of these same undesired r number of input records are removed from the output record $y(t)$ by optimum linear least-squares techniques. To be specific, let the records be ordered so that the first r input records are removed from the remaining $(q - r)$ input records and from the output $y(t)$. The conditioned (residual) records can be denoted by

$$x_{i \cdot r!}(t) = x_{i \cdot 1, 2, \dots, r!}(t)$$
$$y_{y \cdot r!}(t) = y_{y \cdot 1, 2, \dots, r!}(t) \qquad i = r + 1, r + 2, \dots, q$$

From formulas to be derived in Chapter 10, one can compute the conditioned autospectra $S_{ii \cdot r!}$ and $S_{yy \cdot r!}$ for each of these conditioned records, and the conditioned cross-spectra $S_{iy \cdot r!}$ between any pair of these conditioned records. Thus relevant spectral information is available to define a new multiple input/output problem using conditioned records instead of the original arbitrary records.

By definition, the partial coherence function $\gamma_{iy \cdot r!}^2$ is the ordinary coherence function between $x_{i \cdot r!}(t)$ and $y_{y \cdot r!}(t)$ using the conditioned spectral results. Note that when only two inputs are involved, say $x_1(t)$ and $x_2(t)$, then $\gamma_{2y \cdot 1}^2$ is defined by Equation 8.64.

To clarify the relationships between partial and multiple coherence functions, consider again the two input/single output system shown in Figure 8.2. By substituting Equation 8.52 into Equation 8.53 or 8.62, straightforward operations prove that

$$S_{yy \cdot 1, 2} = S_{yy \cdot 1}(1 - \gamma_{2y \cdot 1}^2) \tag{8.67}$$

where

$$\gamma_{2y \cdot 1}^2 = \frac{|S_{2y \cdot 1}|^2}{S_{22 \cdot 1} S_{yy \cdot 1}} \tag{8.68}$$

From Equation 8.54, it follows that

$$S_{yy \cdot 1, 2} = S_{yy}(1 - \gamma_{1y}^2)(1 - \gamma_{2y \cdot 1}^2) \tag{8.69}$$

This represents the extraneous output noise in Figures 8.2 or 8.4, namely

$$S_{nn} = S_{yy \cdot 1, 2} \tag{8.70}$$

The multiple coherence function of Equation 8.35 now becomes

$$\gamma_{y:x}^2 = 1 - (1 - \gamma_{1y}^2)(1 - \gamma_{2y \cdot 1}^2) \tag{8.71}$$

Equation 8.71 is a general result that applies to two input/single output problems where the inputs may be correlated. With uncorrelated inputs, Equation 8.66 shows that

$$\gamma_{y:x}^2 = \gamma_{1y}^2 + \gamma_{2y}^2 \tag{8.72}$$

For the general case of arbitrary inputs, by interchanging the two input records, it is obvious that Equations 8.69 and 8.71 are equivalent to

$$S_{yy\cdot 1,2} = S_{yy}(1 - \gamma_{2y}^2)(1 - \gamma_{1y\cdot 2}^2) \tag{8.73}$$

$$\gamma_{y:x}^2 = 1 - (1 - \gamma_{2y}^2)(1 - \gamma_{1y\cdot 2}^2) \tag{8.74}$$

It follows that

$$(1 - \gamma_{1y}^2)(1 - \gamma_{2y\cdot 1}^2) = (1 - \gamma_{2y}^2)(1 - \gamma_{1y\cdot 2}^2) \tag{8.75}$$

Thus

$$\frac{1 - \gamma_{1y}^2}{1 - \gamma_{2y}^2} = \frac{1 - \gamma_{1y\cdot 2}^2}{1 - \gamma_{2y\cdot 1}^2} \tag{8.76}$$

so that

$$\gamma_{1y}^2 \geq \gamma_{2y}^2 \qquad \text{implies } \gamma_{1y\cdot 2}^2 \geq \gamma_{2y\cdot 1}^2 \tag{8.77}$$

The formula in Equation 8.69 shows that the multiple coherent output spectrum of Equation 8.38 becomes for the two input/single output system

$$\begin{aligned} S_{y:x} = \gamma_{y:x}^2 S_{yy} &= S_{yy} - S_{yy}(1 - \gamma_{1y}^2)(1 - \gamma_{2y\cdot 1}^2) \\ &= \gamma_{1y}^2 S_{yy} + \gamma_{2y\cdot 1}^2 S_{yy\cdot 1} \end{aligned} \tag{8.78}$$

Hence since $S_{nn} = S_{yy\cdot 1,2}$, the total output spectrum

$$\begin{aligned} S_{yy} = \gamma_{y:x}^2 S_{yy} + S_{nn} &= S_{y:x} + S_{nn} \\ &= \gamma_{1y}^2 S_{yy} + \gamma_{2y\cdot 1}^2 S_{yy\cdot 1} + S_{yy\cdot 1,2} \end{aligned} \tag{8.79}$$

By interchanging the two input records,

$$S_{yy} = \gamma_{2y}^2 S_{yy} + \gamma_{1y\cdot 2}^2 S_{yy\cdot 2} + S_{yy\cdot 1,2} \tag{8.80}$$

The results in Equations 8.78 through 8.80 extend systematically to any number of input records. For example, with $q = 3$ inputs,

$$\begin{aligned} S_{yy} = \gamma_{y:x}^2 S_{yy} + S_{nn} \\ = \gamma_{1y}^2 S_{yy} + \gamma_{2y\cdot 1}^2 S_{yy\cdot 1} + \gamma_{3y\cdot 1,2}^2 S_{yy\cdot 1,2} + S_{yy\cdot 1,2,3} \end{aligned} \tag{8.81}$$

The first term in Equation 8.81, namely, $\gamma_{1y}^2 S_{yy}$, is simply the coherent output (power) spectrum for $x_1(t)$ at $y(t)$, as previously defined in Equation 4.30. In terms of one-sided spectral quantities, it is denoted by

$$G_{y:x_1} = G_{y:1} = \gamma_{1y}^2 G_{yy} \tag{8.82}$$

The second term is called the *partial coherent output (power) spectrum* of $x_2(t)$ at $y(t)$ with $x_1(t)$ removed, and is denoted in terms of one-sided spectra by

$$G_{y:x_2\cdot x_1} = G_{y:2\cdot 1} = \gamma_{2y\cdot 1}^2 G_{yy\cdot 1} \tag{8.83}$$

Similarly, the third term gives the partial coherent output spectrum of $x_3(t)$ at $y(t)$ with $x_1(t)$ and $x_2(t)$ removed, that is,

$$G_{y:3 \cdot 1, 2} = \gamma^2_{3y \cdot 1, 2} G_{yy \cdot 1, 2} \tag{8.84}$$

and so on. These quantities can provide valuable physical interpretations in multiple input problems, but the order selected for the input records can strongly influence such interpretations. This is demonstrated later by illustrations in Section 9.4. Further discussions of these matters with appropriate computational schemes are covered in Chapter 10.

REFERENCES

8.1 Bendat, J. S., and Piersol, A. G., *Random Data: Analysis and Measurement Procedures*, Wiley-Interscience, New York, 1971.

8.2 Bendat, J. S., "Solutions for the Multiple Input/Output Problem," *Journal of Sound and Vibration*, Vol. 44, No. 3, p. 311, 1976.

8.3 Bendat, J. S., "System Identification from Multiple Input/Output Data," *Journal of Sound and Vibration*, Vol. 49, No. 3, p. 293, 1976.

CHAPTER 9

ENERGY SOURCE IDENTIFICATION

A major application of the multiple input/output relationships developed in the previous chapter is to source identification problems. For example, assume one is in a factory with a large number of machines that collectively cause substantial acoustic noise at a receiver location of interest. The problem is to determine which machines should be quieted to obtain an optimum reduction in the receiver location noise levels. To do this effectively, it is clear that the contribution of each machine to the receiver noise must be determined.

9.1 PROBLEM FORMULATION

Source identification problems can generally be represented by the model shown in Figure 9.1. Specifically, q number of physical sources producing true inputs $u_i(t)$ are measured by transducers as noise-free signals $w_i(t)$ plus extraneous noise signals $m_i(t)$ to give the measured inputs $x_i(t)$, $i = 1, 2, \ldots, q$. The true inputs $u_i(t)$ pass through constant parameter linear systems $H_{iy}(f)$, $1, 2, \ldots, q$, to produce the true outputs $v_i(t)$, which sum with extraneous noise $n(t)$ to give the total measured output $y(t)$. The frequency response functions $H_{ii}(f)$ relate the true inputs $u_i(t)$ to the noise-free measured inputs $w_i(t)$, and would ideally be $H_{ii}(f) = 1$, $i = 1, 2, \ldots, q$. It is reasonable to assume that the noise terms $m_i(t)$ and $n(t)$ are uncorrelated among themselves and among the inputs $w_i(t)$ and outputs $v_i(t)$. That is,

$$G_{m_i n}(f) = G_{w_i m_i}(f) = G_{w_i n}(f) = G_{v_i n}(f) = 0$$
$$G_{m_i m_j}(f) = G_{w_i m_j}(f) = 0 \qquad i, j = 1, 2, \ldots, q; i \neq j \tag{9.1}$$

9.1.1 *Uncorrelated Sources*

Consider the case where the energy sources $u_i(t)$, $i = 1, 2, \ldots, q$, in Figure 9.1 are uncorrelated, that is,

$$G_{u_i u_j}(f) = 0 \qquad \text{for } i \neq j = 1, 2, \ldots, q \tag{9.2}$$

Assume for the moment that the transducers make accurate, noise-free measurements of the sources so that $x_i(t) = u_i(t)$, that is,

$$H_{ii}(f) = 1 \qquad m_i(t) = 0 \qquad i = 1, 2, \ldots, q \tag{9.3}$$

Under these assumptions, the model in Figure 9.1 is the same as the basic multiple input/output model shown previously in Figure 8.1.

Referring back to Section 3.3.3, for this ideal case, the cross-correlation function between any input $x_i(t)$ and the output $y(t)$ will reveal a single peak (assuming no dispersion), as previously illustrated in Figure 3.13. Furthermore, the contribution of $x_i(t)$ to the total output variance σ_y^2 will be given by the correlated output power relationship

$$\sigma_{y:x_i}^2 = \rho_{iy}^2\!\left(\frac{d}{c}\right)\sigma_y^2 = \sigma_{v_i}^2 \qquad i = 1, 2, \ldots, q \tag{9.4}$$

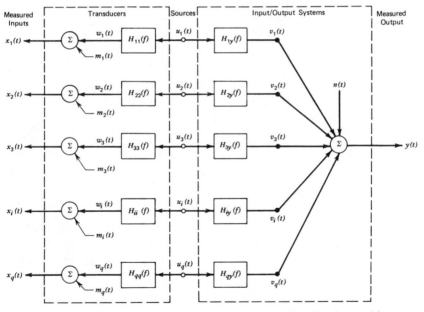

Figure 9.1 Multiple input/output model for source identification problems.

where $\rho_{iy}(d/c)$ is the peak value of the correlation coefficient function between $x_i(t)$ and $y(t)$, as defined in Equation 3.72. However, if the propagation is through r parallel paths, then the correlated output power relationship becomes

$$\sigma^2_{y:x_i} = \sigma^2_y \sum_{k=1}^{r} \rho^2_{iky}\left(\frac{d_k}{c_k}\right) = \sigma^2_{v_i} \qquad i = 1, 2, \ldots, q \qquad (9.5)$$

where $\rho_{iky}(d_k/c_k)$, $k = 1, 2, \ldots, r$, are the peak values of the correlation coefficient function between $x_i(t)$ and $y(t)$ corresponding to the various paths, as discussed in Section 6.1. Finally, if one or more of the propagation paths is frequency dispersive, the application of Equation 9.5 is further complicated by the practical considerations discussed in Section 6.3.

If the only interest is in the contribution of $x_i(t)$ to $y(t)$, independent of possible propagation paths and dispersion, the most effective way to approach the problem is in the frequency domain using the coherent output power relationship of Equation 4.30. Specifically, for the situation of q uncorrelated inputs, the autospectrum of $y(t)$ due to the input $x_i(t)$ alone is given by

$$G_{y:x_i}(f) = \gamma^2_{x_iy}(f)G_{yy}(f) = G_{v_iv_i}(f) \qquad i = 1, 2, \ldots, q \qquad (9.6)$$

where $\gamma^2_{x_iy}(f)$ is the ordinary coherence function between $x_i(t)$ and $y(t)$, previously defined in Equation 3.43. It follows that the total autospectrum of $y(t)$ due to all measured inputs will be

$$G_{y:x}(f) = G_{yy}(f) \sum_{i=1}^{q} \gamma^2_{x_iy}(f) = G_{vv}(f) \qquad (9.7)$$

The sum of ordinary coherence terms in Equation 9.7 is the multiple coherence function for the uncorrelated input case, as demonstrated in Section 8.2.

9.1.2   Noise-Free Source Measurements

Now consider the case where the true source inputs $u_i(t)$, $i = 1, \ldots, q$, are uncorrelated, as given by Equation 9.2, but the frequency response functions between the actual source inputs and the measured inputs is not unity as in Equation 9.3, that is, $u_i(t) \neq x_i(t)$. Still assuming the input measurement noise $m_i(t) = 0$ so that $x_i(t) = w_i(t)$ for all i, it follows from Equation 4.8 that

$$G_{x_ix_i}(f) = |H_{ii}(f)|^2 G_{u_iu_i}(f) \qquad i = 1, 2, \ldots, q \qquad (9.8)$$

and the cross-spectrum between the measured input and output becomes

$$G_{x_iy}(f) = H_{ii}^*(f)G_{u_iy}(f) = H_{ii}^*(f)H_{iy}(f)G_{u_iu_i}(f) \qquad (9.9)$$

The measured input/output coherence function is now given by

$$\gamma_{x_iy}^2(f) = \frac{|G_{x_iy}(f)|^2}{G_{x_ix_i}(f)G_{yy}(f)} = \frac{|H_{iy}(f)|^2 G_{u_iu_i}(f)}{G_{yy}(f)} \qquad (9.10)$$

But

$$|H_{iy}(f)|\,^2G_{u_iu_i}(f) = G_{v_iv_i}(f) \qquad (9.11)$$

Hence

$$G_{y:x_i}(f) = \gamma_{x_iy}^2(f)G_{yy}(f) = G_{v_iv_i}(f) \qquad i = 1, 2, \ldots, q \qquad (9.12)$$

exactly as stated before in Equation 9.6.

The result in Equation 9.12 demonstrates an important fact, namely, any measurement $x_i(t)$ that is noise-free and linearly related to the true source $u_i(t)$ will produce a correct result when used in coherent output power calculations. In practice, this means that any convenient transducer can be used to measure the energy source as long as the transducer output is linearly related to the energy. It is not even necessary to calibrate the transducer; all calibrations are effectively accomplished by the coherence function calculation.

9.1.3 *Practical Considerations*

The procedure for identifying energy sources using Equation 9.6 is straightforward in theory, but numerous practical problems can severly limit its applications. It is for this reason that an analysis using Equation 9.6 should be attempted only when other more direct source identification procedures are not feasible. For example, if all the sources except the one of interest can be physically turned off, then obviously this should be done so that the source contribution to $y(t)$ can be measured directly. Perhaps the other sources cannot be turned off, but the source of interest can be. If so, the contribution of the other sources to $y(t)$ could be measured directly and subtracted from the overall at $y(t)$ to obtain the contribution of the source of interest. In those cases where the inputs of interest are narrow band in character, or better yet periodic, it may be possible to identify the contribution of each source to $y(t)$ by isolating and measuring their corresponding narrow-band components in $y(t)$, assuming that two or more inputs do not have components which overlap in frequency. An analysis using Equation 9.6 is generally appropriate only when the various sources must occur simultaneously and are random covering a common frequency range.

The types of problems that limit the application of Equation 9.6 in practice as as follows:

1. Inaccurate source measurements.
2. Input measurement noise.
3. Input/output time delay errors.
4. Reverberation effects.
5. Input/output feedback.
6. Input/output nonlinearities.
7. Source measurement interference.
8. Correlation among sources.
9. Common source periodicities.
10. Statistical sampling errors.

The first six items are single channel problems that can arise whether or not the input measurements are correlated. These matters are discussed in Section 9.2. The next three items are related to problems due to correlation among the input measurements. Such correlation can occur for two reasons: (*a*) due to interference (cross-talk) in the input measurements or (*b*) because of physical correlation among the sources. These matters are discussed in Sections 9.3 and 9.4, respectively. Statistical sampling errors are summarized in Section 9.4.3. and detailed in Chapter 11.

9.2 SINGLE CHANNEL MEASUREMENT PROBLEMS

Even when the sources of interest in a multiple source problem are un-correlated, a number of practical problems can limit the application of the coherent output power calculation of Equation 9.6. Such potential problems must be carefully considered prior to an experiment.

9.2.1 *Input Measurement Problems*

Referring to Figure 9.1, it is assumed that all inputs $x_i(t)$, $i = 1, 2, \ldots, q$, representing potential sources of interest can be measured. In practical terms, this suggests that one can (*a*) accurately guess the locations of potentially significant sources, (*b*) reach them with a transducer, and (*c*) devise an appropriate transducer that will accurately measure them. The first two requirements are sometimes difficult to meet, either because one has no idea where the significant sources might be or because they are impossible to reach with transducers. In either case, an analysis procedure based on receiver measurements only must be used, as detailed in Chapter 7. In other cases, the first two requirements can be complied with, but the third poses a problem. This is particularly true of distributed sources, for example, acoustic sources of aerodynamic origin such as the noise made by air rushing through a pipe. It is usually necessary to measure such a source with multiple transducers as if it were a collection of sources.

Beyond the physical problems of measuring suspected sources, one must be careful about extraneous measurement noise. As discussed in Section 4.2, the coherence function in Equation 9.6 will be suppressed in value by the presence of extraneous noise in either the input or output measurements. However, the interpretations of the results needed to identify source contributions assume that all extraneous noise, if any, is at the output location only; the input measurements $x_i(t)$, $i = 1, 2, \ldots, q$, are assumed to be noise free. If there is extraneous noise in the input measurements, this will reduce the calculated coherence functions and cause Equation 9.6 to underestimate the contribution of each measured source to the output $y(t)$. Specifically, if the input measurement $x_i(t) = w_i(t) + m_i(t)$ where $w_i(t)$ is the true input and $m_i(t)$ is measurement noise, then from Equation 4.48 the indicated coherent output power will be

$$\hat{G}_{y:x_i}(f) = G_{y:w_i}(f) \frac{G_{w_iw_i}(f)}{G_{w_iw_i}(f) + G_{m_im_i}(f)} \tag{9.13}$$

Hence every effort should be made to have the maximum signal-to-noise ratio feasible at the input measurements.

9.2.2 Time Delay Bias Errors

It is common practice in experiments involving multiple time history measurements to collect all records on a common time base. From the viewpoint of coherence analysis, this is acceptable as long as the record length T is very long compared to the time displacement τ_1 between any two records. However, as $\tau_1 \rightarrow T$, the estimated coherence between the records will diminish due to a time delay bias error, shown in Equation 11.31 of Chapter 11 as

$$\hat{\gamma}_{xy}^2(f) \approx \left(1 - \frac{\tau_1}{T}\right)^2 \gamma_{xy}^2(f) \tag{9.14}$$

where $\hat{\gamma}_{xy}^2(f)$ is the estimate of $\gamma_{xy}^2(f)$. The result in Equation 9.14 is exact only for the case where $x(t)$ and $y(t)$ are theoretical white noise producing delta-function-type correlations. However, Equation 9.14 is a good approximation in practice as long as the bandwidths of $x(t)$ and $y(t)$ are such that $BT \gg 1$.

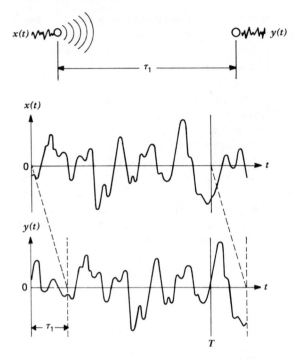

Figure 9.2 Propagation time delay problem.

The time delay bias error in Equation 9.14 will always occur in any problem where there is a physical propagation from the input $x(t)$ to the output $y(t)$ and the data are collected on a common time base, as illustrated in Figure 9.2. The error can be particularly severe when coherence functions are calculated by modern FFT (fast Fourier transform) operations, because these procedures generally compute coherence functions from spectral estimates which are calculated by ensemble averaging the results of many repeated analyses of relatively short records, as described in Section 3.4.2. However, the error is easily avoided by estimating the propagation times anticipated in the data and introducing a precomputational delay into the analysis to correct for the anticipated physical delay. Most of the modern analyzers are equipped with appropriate delay circuits for this purpose.

9.2.3 Reverberation Effects

Two reverberation situations are of concern. The first is where both the input $x(t)$ and the output $y(t)$ are exposed to multiple reflections, as discussed earlier in Section 7.3.4. It is noted in that section that such reverberation under certain conditions can cause the measured coherence function between $x(t)$ and $y(t)$ to approach the form of diffuse noise given in Equation 7.64. This would cause Equation 9.6 to underestimate source contributions and constitutes a serious limitation on the source identification procedures. The second situation is where the reverberation occurs at the output only and does not significantly influence $x(t)$. This case can be viewed as a problem of multiple paths between $x(t)$ and $y(t)$, each with a different propagation time, as illustrated in Figure 9.3 and discussed in Chapter 6. A precomputational delay procedure would be useful here to correct for the propagation time along the direct path, but the bias error in Equation 9.14 will still be involved in all of the reflected paths. Hence if a correct coherence function is to be measured, the analysis must be performed with a record length T that is very long compared to the propagation time of all significant reflections. It should be emphasized here that the record length T of concern is the length of each data block used for the spectral estimates and not the total length of the analyzed data given in Section 3.4.2 by $T_{\text{total}} = n_d T$.

The magnitude of the time delay bias error in problems involving reverberation at the output is sometimes assessed in terms of the ratio T_r/T, where T_r

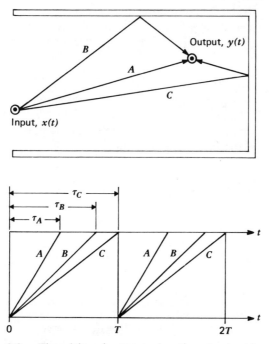

Figure 9.3   Time delays due to reverberation at output location.

is the reverberation time defined as follows: If the energy source producing $x(t)$ is instantaneously terminated, T_r is the time required for the variance of $y(t)$ to fall 60 dB ($10^6:1$). Empirical data for the bias error in coherence measurements between a reflection-free acoustic noise source and a measured output inside a reverberant enclosure are shown in Figure 9.4. This experiment covered a frequency range of 0.1 to 10 kHz in a room with a volume of about 60 meters^2. The results in Figure 9.4 indicate that the coherence analysis must be performed with a record length of $T > T_r$ if significant time delay bias errors are to be avoided.

As a concluding point on this subject, it should be mentioned that the time delay bias error can sometimes be used to advantage in analyzing data with output reverberation. Specifically, there may be cases where only the direct path contribution of an input to an output is of interest. If the source has a sufficiently broad bandwidth so that the peaks in $R_{xy}(\tau)$ corresponding to the direct path and first reflection are well separated, then this can be accomplished using a properly selected record length for the analysis. Specifically, assume that the cross-correlation function between $x(t)$ and $y(t)$ has a form similar to that defined in Equation 6.2. Let τ_1 and τ_2 be the two time delays of the correlation peaks due to the direct path and the first reflected path, respectively, where the time delays for all other reflections are $\tau_k > \tau_2$, $k = 3, 4, 5, \ldots$. If the coherence function between $x(t)$ and $y(t)$ is now computed with a record length $\tau_1 < T < \tau_2$, the resulting biased estimate $\hat{\gamma}_{xy}^2(f, T)$ will represent only the direct path contribution of $x(t)$ to $y(t)$, that is,

$$G_{y:x_d}(f) = \hat{\gamma}_{xy}^2(f, T)G_{yy}(f) \qquad \tau_1 < T < \tau_2 \qquad (9.15)$$

To illustrate this procedure, consider an acoustic experiment performed by Halvorsen [9.1], as outlined previously in Figure 9.3. With broad-band noise as the source $x(t)$, the autospectrum of the output $y(t)$ was measured for the direct path only using Equation 9.15. The enclosure about $y(t)$

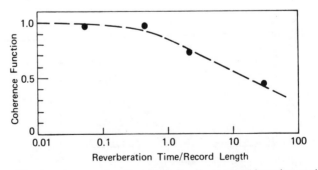

Figure 9.4 Bias in coherence function estimates due to reverberation at the output.

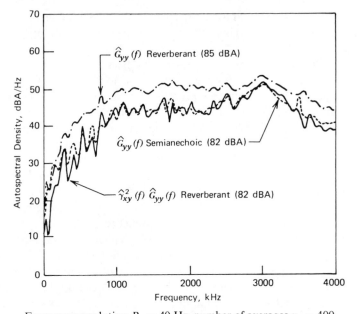

Frequency resolution $B_e = 40$ Hz, number of averages $n_d = 400$

Figure 9.5 Biased coherent output power calculation with reverberation at the output. (These data are taken from Reference 9.1 with the permission of Acoustical Publications, Inc. and the authors.)

was then removed to obtain a semi-anechoic situation, and the autospectrum of $y(t)$ was again measured. All calculations were performed using a resolution of $B_e = 40$ Hz and $n_d = 400$ averages. The results are presented in Figure 9.5. Note that the direct path contribution calculated from Equation 9.15 is in excellent agreement with the semi-anechoic results at most frequencies.

9.2.4. Input/Output Feedback and Nonlinearities

The result in Equation 9.6 assumes an open loop from $x(t)$ to $y(t)$, that is, the input $x(t)$ causes an output $y(t)$ that cannot feedback to influence $x(t)$. The analysis of feedback problems requires additional measurements as developed in Section 4.3. There are a number of situations in practice where input/output feedback might occur. The reverberation problem discussed in the previous section, where $x(t)$ and $y(t)$ are exposed to reflections, essentially constitutes a feedback problem. As a more subtle illustration, assume that an intense acoustic source causes a structural vibration of interest. Assume also that the source is monitored by a microphone and the structural vibration by an accelerometer. The structural vibration, however, will radiate acoustic energy that might be sensed by the microphone if it is located too close to the structure. This situation constitutes feedback and will lead to erroneous results.

Concerning nonlinearities, it is shown in Section 5.1.1 that the basic formulation leading to Equation 9.6 makes all nonlinearities between $x(t)$ and $y(t)$ appear as statistically independent (extraneous) noise at the output, that is, nonlinear effects appear as part of $n(t)$ in Figure 9.1. Nonlinearities will therefore cause Equation 9.6 to underestimate the contribution of a measured source. Hence it is important that one has reason to believe the propagation paths in a source identification problem are linear.

In mechanical vibration problems, a common type of nonlinearity is when a vibration source excites that nonlinear mechanism which is usually referred to as a "rattle." If the rattle is close to the output location so as to dominate $y(t)$, then the coherence function between the actual vibration source measured by $x(t)$ and the output $y(t)$ will be strongly suppressed, indicating that the actual source is not a significant contributor. Conversely, if a measurement were made near the location of the rattle, it would be identified as the source location. In some cases, this may be appropriate if the interest is specifically in finding rattles rather than vibration sources. In other cases, however, rattles or other nonlinearities will only confuse the analysis and perhaps lead to serious misinterpretations.

9.3   INPUT MEASUREMENT INTERFERENCE

Referring back to Figure 9.1, it is assumed that the input measurement $x_i(t)$ will sense some linear function of the true source input $u_i(t)$, but will not sense any other source $u_j(t)$, $i \neq j$. In practice, it is sometimes difficult to design a transducer that will measure an individual source without sensing other nearby sources as well. This problem is particularly severe in acoustic noise work where microphones are commonly used to measure input sources. Such interference will cause the input measurements $x_i(t)$, $i = 1, 2, \ldots, q$, to be correlated even though the actual sources $u_i(t)$, $i = 1, 2, \ldots, q$, may be uncorrelated. In such cases, Equation 9.6 will overestimate the contributions of the individual sources.

9.3.1   Coherent Output Power Calculations

To demonstrate the input measurement interference problem, consider the case of two statistically independent sources $u_1(t)$ and $u_2(t)$ that produce an output $y(t)$, as shown in Figure 9.6. Source 1 is measured by transducer 1 through a frequency response function $H_{11}(f)$, and Source 2 is measured by transducer 2 through a frequency response function $H_{22}(f)$. However,

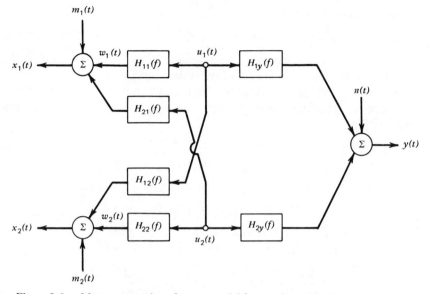

Figure 9.6 Measurement interference model for two input/single output system.

transducer 2 also detects some of source 1 through $H_{12}(f)$, and transducer 1 picks up some of source 2 through $H_{21}(f)$. The sources transmit to the output through $H_{1y}(f)$ and $H_{2y}(f)$, respectively. There may be extraneous noise $n(t)$ at the output, but it is assumed that there is no extraneous noise at the inputs. To obtain quantitative results, assume that the frequency response functions at a given frequency are

$$H_{11} = H_{22} = 1.00 \qquad H_{1y} = H_{2y} = 0.50$$

Further assume that the true source inputs $u_1(t)$ and $u_2(t)$ have autospectra at the same frequency given by

$$G_{u_1u_1} = 1.00A \qquad G_{u_2u_2} = 2.00A$$

First consider the ideal case where there is no extraneous noise at the inputs or the output and no input measurement interference, that is, $H_{12} = H_{21} = 0$. It follows from material in Section 8.1.3 that

$$G_{x_1x_1} = G_{u_1u_1} = 1.00A \qquad G_{x_2x_2} = G_{u_2u_2} = 2.00A$$

$$G_{x_1y} = G_{u_1y} = H_{1y}G_{u_1u_1} = 0.50A \qquad G_{x_2y} = G_{u_2y} = H_{2y}G_{u_2u_2} = 1.00A$$

$$G_{yy} = |H_{1y}|^2G_{u_1u_1} + |H_{2y}|^2G_{u_2u_2} = 0.75A$$

$$\gamma^2_{x_1y} = \frac{|G_{x_1y}|^2}{G_{x_1x_1}G_{yy}} = 0.333 \qquad \gamma^2_{x_2y} = \frac{|G_{x_2y}|^2}{G_{x_2x_2}G_{yy}} = 0.667$$

Hence Equation 9.6 yields

$$G_{y:x_1} = \gamma_{x_1y}^2 G_{yy} = 0.25A \qquad G_{y:x_2} = \gamma_{x_2y}^2 G_{yy} = 0.50A \qquad (9.16)$$

which is obviously the correct answer; since $H_{1y} = H_{2y}$, one-third of the output must be due to source 1 and two-thirds due to source 2.

Now consider the case where there is still no extraneous noise, but input measurement interference occurs such that $H_{12} = H_{21} = 0.50$. Measured results will be

$$G_{x_1x_1} = |H_{11}|^2 G_{u_1u_1} + |H_{21}|^2 G_{u_2u_2} = 1.50A$$

$$G_{x_2x_2} = |H_{12}|^2 G_{u_1u_1} + |H_{22}|^2 G_{u_2u_2} = 2.25A$$

$$G_{x_1y} = H_{11}^* H_{1y} G_{u_1u_1} + H_{21}^* H_{2y} G_{u_2u_2} = 1.00A$$

$$G_{x_2y} = H_{12}^* H_{1y} G_{u_1u_1} + H_{22}^* H_{2y} G_{u_2u_2} = 1.25A$$

$$G_{x_1x_2} = H_{11}^* H_{12} G_{u_1u_1} + H_{21}^* H_{22} G_{u_2u_2} = 1.50A$$

$$G_{yy} = |H_{1y}|^2 G_{u_1u_1} + |H_{2y}|^2 G_{u_2u_2} = 0.75A$$

$$\gamma_{x_1y}^2 = \frac{|G_{x_1y}|^2}{G_{x_1x_1} G_{yy}} = 0.889 \qquad \gamma_{x_2y}^2 = \frac{|G_{x_2y}|^2}{G_{x_2x_2} G_{yy}} = 0.926$$

Thus Equation 9.6 gives

$$G_{y:x_1} = \gamma_{x_1y}^2 G_{yy} = 0.667A \qquad G_{y:x_2} = \gamma_{x_2y}^2 G_{yy} = 0.694A \qquad (9.17)$$

The discrepancy between the results in Equations 9.16 and 9.17 represents the effects of source measurement interference. Note that the contribution of each source is exaggerated and the relative difference between the sources is understated.

Finally, consider the case where there is extraneous noise at the output with an autospectrum of $G_{nn} = 0.25A$ as well as input measurement interference with $H_{12} = H_{21} = 0.5$ as before. For this case, the input autospectra and the input/output cross-spectra terms are the same as those leading to Equation 9.17. The output autospectrum and coherence terms become

$$G_{yy} = |H_{1y}|^2 G_{u_1u_1} + |H_{2y}|^2 G_{u_2u_2} + G_{nn} = 1.00A$$

$$\gamma_{x_1y}^2 = \frac{|G_{x_1y}|^2}{G_{x_1x_1} G_{y_1y_1}} = 0.667 \qquad \gamma_{x_2y}^2 = \frac{|G_{x_2y}|^2}{G_{x_2x_2} G_{yy}} = 0.694$$

These results are different than before. However, Equation 9.6 yields

$$G_{y:x_1} = \gamma_{x_1y}^2 G_{yy} = 0.667A \qquad G_{y:x_2} = \gamma_{x_2y}^2 G_{yy} = 0.694A \qquad (9.18)$$

which is the same result obtained with no extraneous noise at the output in Equation 9.17, as it should be. Extraneous noise in $y(t)$ alters the fractional portion of $y(t)$ due to either $x(t)$, but does not change the value of their absolute contributions.

9.3.2 *Partial Coherence Function Calculations*

The question now arises as to how partial coherence functions will assist in evaluating problems when input measurement interference is present. To help answer this question, consider again the two input illustration outlined in Figure 9.6. From Section 8.3, the partial coherence function between each input and the output is given by

$$\gamma^2_{x_i y \cdot x_j} = \frac{|G_{x_i y \cdot x_j}|^2}{G_{x_i x_i \cdot x_j} G_{yy \cdot x_j}} \qquad i \neq j = 1, 2 \qquad (9.19)$$

where the conditioned spectral terms are

$$G_{x_1 x_1 \cdot x_2} = G_{x_1 x_1}(1 - \gamma^2_{x_1 x_2}) \qquad G_{x_2 x_2 \cdot x_1} = G_{x_2 x_2}(1 - \gamma^2_{x_2 x_1})$$

$$G_{x_1 y \cdot x_2} = G_{x_1 y} - \frac{G_{x_1 x_2} G_{x_2 y}}{G_{x_2 x_2}} \qquad G_{x_2 y \cdot x_1} = G_{x_2 y} - \frac{G_{x_2 x_1} G_{x_1 y}}{G_{x_1 x_1}}$$

$$G_{yy \cdot x_2} = G_{yy}(1 - \gamma^2_{x_2 y}) \qquad G_{yy \cdot x_1} = G_{yy}(1 - \gamma^2_{x_1 y})$$

For the first case in Section 9.3.1, where there is no input measurement inference ($H_{12} = H_{21} = 0$), it follows from the above relationships and the basic quantities calculated in that section that

$$\begin{aligned} G_{x_1 x_1 \cdot x_2} &= 1.00A & G_{x_2 x_2 \cdot x_1} &= 2.00A \\ G_{x_1 y \cdot x_2} &= 0.50A & G_{x_2 y \cdot x_1} &= 1.00A \\ G_{yy \cdot x_2} &= 0.25A & G_{yy \cdot x_1} &= 0.50A \end{aligned}$$

Hence from Equation 9.19, the partial coherence values are

$$\gamma^2_{x_1 y \cdot x_2} = 1.00 \qquad \gamma^2_{x_2 y \cdot x_1} = 1.00 \qquad (9.20)$$

demonstrating that when one input is conditioned out of the calculations, there is a perfect linear relationship between the second input and the output, as would be expected.

For the second case, where the interference is represented by $H_{12} = H_{21} = 0.50$,

$$G_{x_1 x_2} = H^*_{11} H_{12} G_{u_1 u_1} + H^*_{21} H_{22} G_{u_2 u_2} = 1.50A$$

and the following results are obtained:

$$\begin{aligned} G_{x_1 x_1 \cdot x_2} &= 0.50A & G_{x_2 x_2 \cdot x_1} &= 0.75A \\ G_{x_1 y \cdot x_2} &= 0.167A & G_{x_2 y \cdot x_1} &= 0.250A \\ G_{yy \cdot x_2} &= 0.0555A & G_{yy \cdot x_1} &= 0.0833A \\ \gamma^2_{x_1 y \cdot x_2} &= 1.00 & \gamma^2_{x_2 y \cdot x_1} &= 1.00 \end{aligned} \qquad (9.21)$$

Hence measurement interference does not influence the partial coherence between a given input and the output; it still reveals a perfect linear relationship.

Finally, when extraneous noise is added at the output with an auto-spectrum $G_{nn} = 0.25A$, the results become

$$
\begin{aligned}
G_{x_1 x_1 \cdot x_2} &= 0.50A & G_{x_2 x_2 \cdot x_1} &= 0.75A \\
G_{x_1 y \cdot x_2} &= 0.167A & G_{x_2 y \cdot x_1} &= 0.250A \\
G_{yy \cdot x_2} &= 0.30A & G_{yy \cdot x_1} &= 0.33A \\
\gamma^2_{x_1 y \cdot x_2} &= 0.19 & \gamma^2_{x_2 y \cdot x_1} &= 0.25
\end{aligned}
\tag{9.22}
$$

The results in Equation 9.22 are interpreted as follows. When measurement $x_2(t)$ is conditioned out of the analysis, 19% of the output can be attributed to input $x_1(t)$ and the remaining 81% to extraneous noise. Conversely, when $x_1(t)$ is conditioned out of the analysis, 25% of the output can be attributed to $x_2(t)$ and 75% to extraneous noise. It is important to note that the partial coherence functions *do not* yield the fractional portion of the output due to one input relative to the other, as might be desired in practice. However, the measurement of the more intense source does yield the larger partial coherence. Assuming that the interference is similar among all transducers, this will generally be true and constitutes an important qualitative interpretation of the partial coherence in such problems.

9.3.3 Suppression of Measurement Interference

The only way to suppress the measurement interference problem is to measure the source inputs $u_i(t)$, $i = 1, 2, \ldots, q$, with transducers that are as insensitive as feasible to sources other than the specific sources of interest. The results developed in Section 9.1.2 are particularly helpful here, namely, any transducer can be used as long as it produces a signal that is linearly related to the source energy. For example, if an acoustic source is due to radiation from a vibrating structure, then the source can be measured using a motion transducer (an accelerometer) on the structure rather than a pressure transducer (a microphone) near the structure. Since the motion of the structure has a direct linear relationship with the acoustic noise it radiates, a motion transducer at one point will produce an acceptable measure of the noise as long as the spatial coherence of the structural vibration is relatively strong over the radiating surface. Of course, the motion measurement will not be completely free of interference since it will sense the structural vibration excited by other acoustic sources. In many cases, however, this contaminating vibration will be substantially less than the interference in a pressure measurement.

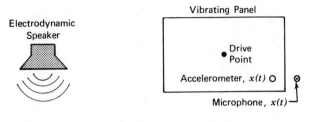

Electrodynamic
Speaker

Vibrating Panel

• Drive
Point

Accelerometer, $x(t)$ O   ⊗

Microphone, $x(t)$

⊗ Microphone, $y(t)$

Figure 9.7   Acoustically radiating panel experiment.

To demonstrate the above points, consider the experiment outlined in Figure 9.7 where a panel section excited by broad-band random vibration radiates acoustic noise to an output microphone. The autospectrum of the noise radiated by the panel, as seen at the output with no background noise, is shown in Figure 9.8. Now, intense statistically independent background noise is added using a speaker near the panel to produce acoustic energy that is about 27 dB higher (500 times greater) than the radiated noise from the panel. The autospectrum of the output for this case is also shown in Figure 9.8. When the panel noise is measured by an input microphone near the panel, the resulting coherent output power calculated using Equation 9.6

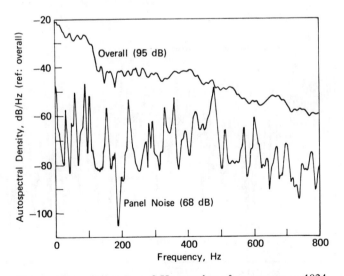

Frequency resolution $B_e = 2$ Hz, number of averages $n_d = 1024$

Figure 9.8   Autospectra of output noise from panel alone and from panel plus background noise.

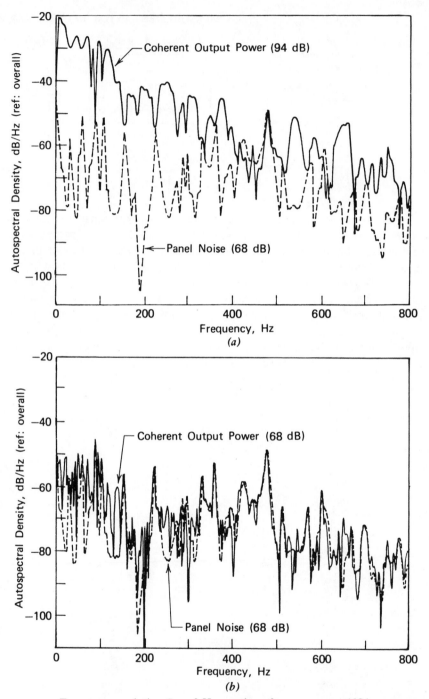

Frequency resolution $B_e = 2$ Hz, number of averages $n_d = 1024$

Figure 9.9 Coherent output power calculations for panel-radiated noise. (a) Using microphone for input measurement. (b) Using accelerometer for input measurement.

is totally incorrect, as seen in Figure 9.9*a*. This is because the input micro-phone hears far more of the background noise than it does the source. On the other hand, when the panel-radiated noise is measured by a single ac-celerometer on the panel, the resulting coherent output power calculation produces a reasonably accurate estimate of the radiated noise at the output location, as shown in Figure 9.9*b*. All data in Figures 9.8 and 9.9 were computed with a resolution of $B_e = 2$ Hz and $n_d = 1024$ averages.

9.4 PHYSICALLY CORRELATED SOURCES

The final problem of interest in source identification procedures is where two or more of the sources of interest are physically correlated. Here the situa-tion differs from the input measurement interference problem discussed in the previous section in that it is now the actual source inputs $u_i(t)$, and not just their measurements $x_i(t)$, $i = 1, 2, \ldots, q$, that are correlated. It is assumed here, however, that no two sources are perfectly correlated since this cannot be resolved, as discussed in Section 8.1.3. Note that two sine waves at the same frequency are always perfectly correlated. Hence these procedures are not applicable to problems involving periodic sources with common frequencies.

9.4.1 *Coherent Output Power Analysis*

From the developments in Chapter 8, it is clear that the coherent output power calculation of Equation 9.6 will always exaggerate the contribution of individual sources when correlation exists among the sources. To demon-strate and clarify this fact, consider the two-source illustration in Figure 9.10. This situation is similar to that previously illustrated for the input measurement interference problem in Figure 9.6, except now the sources

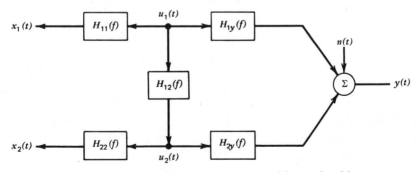

Figure 9.10 Two input/single output system with correlated inputs.

are correlated without any measurement interference. Specifically, assume that source 2 is due in part to source 1 acting through the frequency response function H_{12}, where $U_2 = U_{2 \cdot 1} + H_{12} U_1$.

To obtain quantitative results, as in Section 9.3.1, let

$$H_{12} = H_{11} = H_{22} = 1.00 \qquad H_{1y} = H_{2y} = 0.50$$

Further assume that the *statistically independent* part of the source inputs are $1.00A$ for both sources 1 and 2. It follows that the autospectra of the true source inputs and their measurements will be given by

$$G_{x_1 x_1} = G_{u_1 u_1} = G_{11} = 1.00A \qquad G_{x_2 x_2} = G_{u_2 u_2} = G_{22} = 2.00A$$

$$G_{x_1 x_2} = G_{u_1 u_2} = G_{12} = 1.00A \qquad \gamma_{12}^2 = \frac{|G_{12}|^2}{G_{11} G_{12}} = 0.50$$

First, for the case of no extraneous noise at the output, it follows from material in Section 8.1.3 that

$$G_{1y} = H_{1y} G_{11} + H_{2y} G_{12} = 1.00A \qquad G_{2y} = H_{2y} G_{22} + H_{1y} G_{21} = 1.50A$$

$$G_{yy} = |H_{1y}|^2 G_{11} + H_{1y}^* H_{2y} G_{12} + H_{2y}^* H_{1y} G_{21} + |H_{2y}|^2 G_{22} = 1.25A$$

$$\gamma_{1y}^2 = \frac{|G_{1y}|^2}{G_{11} G_{yy}} = 0.80 \qquad \gamma_{2y}^2 = \frac{|G_{2y}|^2}{G_{22} G_{yy}} = 0.90$$

Hence the coherent output powers from Equation 9.6 are

$$G_{y:1} = \gamma_{1y}^2 G_{yy} = 1.00A \qquad G_{y:2} = \gamma_{2y}^2 G_{yy} = 1.125A \qquad (9.23)$$

These results are obviously incorrect when viewed in total because $G_{y:1} + G_{y:2} > G_{yy}$. However, if one correctly deduced that the correlation between the sources physically originated at source 1, then the $G_{y:1}$ result in Equation 9.23 would have an accurate interpretation. Specifically, $G_{y:1} = 1.00A$ is the total contribution to $y(t)$ that can be attributed to source 1 *through all possible paths* to $y(t)$. This follows since the statistically independent contribution of source 2 to $y(t)$ is

$$G_{y:2 \cdot 1} = |H_{2y}|^2 G_{22 \cdot 1} = |H_{2y}|^2 A = 0.25A \qquad (9.24)$$

Hence the total contribution of source 1 including all correlated effects with source 2 is

$$G_{y:1} = G_{yy} - G_{y:2 \cdot 1} = 1.00A \qquad (9.25)$$

as given in Equation 9.23. From another viewpoint, if source 2 were turned off but still transmitted energy to the output due to the correlation physically originating at source 1, it follows from Equation 8.6 that

$$G_{yy} = |H_{1y} + H_{2y} H_{12}|^2 G_{11} = 1.00A \qquad (9.26)$$

Of course, if one incorrectly assumed that the correlation between the sources physically originated at source 2 rather than source 1, then these interpretations of the results in Equation 9.23 would be incorrect and seriously misleading.

Now consider the case where there is extraneous noise at the output with an autospectrum of $G_{nn} = 0.25A$. The total output now becomes $G_{yy} = 1.50A$ and

$$\gamma_{1y}^2(f) = 0.667 \qquad \gamma_{2y}^2(f) = 0.750$$

It follows from Equation 9.6 that

$$G_{y:1} = 1.00A \qquad G_{y:2} = 1.125A \qquad (9.27)$$

exactly as for the noise-free case in Equation 9.23. The presence of extraneous noise at the output does not influence the coherent output power calculations except to add random error to the results, as will be discussed later.

9.4.2   Partial Coherent Output Power Analysis

From the developments in Chapter 8, the problem posed by correlated sources can be somewhat (but not totally) resolved using partial coherent output calculations, as given for the two input case by Equation 8.83. To help clarify these applications, again consider the illustration in Figure 9.10. First, assume that there is no extraneous noise at the output; that is, $n(t) = 0$. From the relationships in Sections 8.3 and 9.4.1,

$$G_{1y\cdot 2} = G_{1y} - \frac{G_{12}G_{2y}}{G_{22}} = 0.25A \qquad G_{2y\cdot 1} = G_{2y} - \frac{G_{21}G_{1y}}{G_{11}} = 0.50A$$

$$G_{11\cdot 2} = G_{11}(1 - \gamma_{12}^2) = 0.50A \qquad G_{22\cdot 1} = G_{22}(1 - \gamma_{21}^2) = 1.00A$$

$$G_{yy\cdot 2} = G_{yy}(1 - \gamma_{2y}^2) = 0.125A \qquad G_{yy\cdot 1} = G_{yy}(1 - \gamma_{1y}^2) = 0.25A$$

$$\gamma_{1y\cdot 2}^2 = \frac{|G_{1y\cdot 2}|^2}{G_{11\cdot 2}G_{yy\cdot 2}} = 1.00 \qquad \gamma_{2y\cdot 1}^2 = \frac{|G_{2y\cdot 1}|^2}{G_{22\cdot 1}G_{yy\cdot 1}} = 1.00$$

Hence the partial coherent output powers are given by Equation 8.83 as

$$G_{y:1\cdot 2} = \gamma_{1y\cdot 2}^2 G_{yy\cdot 2} = 0.125A \qquad G_{y:2\cdot 1} = \gamma_{2y\cdot 1}^2 G_{yy\cdot 1} = 0.25A \quad (9.28)$$

Unlike the coherent output powers in Section 9.4.1, the partial coherent output power calculations yield a correct interpretation for the contribution of source 2, but not source 1. Specifically, from Equation 9.24, the statistically independent contribution of source 2 is $G_{y:2\cdot 1} = 0.25A$, exactly as calculated in Equation 9.28. On the other hand, the value $G_{y:1\cdot 2}$ in Equation 9.28 is not physically meaningful. The presence of extraneous noise at the output will not alter these results and conclusions.

In summary, if one correctly deduced that the correlation between the sources in Figure 9.10 physically originated at source 1 as shown, then the contribution of the sources to $y(t)$ would be accurately given by Equation 8.78, namely

$$G_{y:x} = \gamma^2_{1y}G_{yy} + \gamma^2_{2y\cdot1}G_{yy\cdot1} \tag{9.29}$$

where

$$\gamma^2_{1y}G_{yy} = 1.00A = \text{contribution of source 1}$$
$$\gamma^2_{2y\cdot1}G_{yy\cdot1} = 0.25A = \text{contribution of source 2}$$

Note that the above results are not just mathematically correct, but are physically correct also. If source 1 were turned off, the output spectrum would in fact drop by $1.00A$. If source 2 were turned off, the output spectrum would drop by $0.25A$, assuming that source 2 still radiated the energy transmitted from source 1. In practice, of course, the problem is to correctly deduce the physical origin of the correlation among the sources.

To illustrate these ideas further, consider the laboratory vibroacoustic experiment outlined in Figure 9.11. This experiment was performed by Barrett [9.2] and involved an instrument panel mounted on a truss structure. The instrument panel was excited by highly correlated mechanical vibration at each end of the truss plus statistically independent acoustic noise over a frequency range from 5 to 1000 Hz. Two accelerometers mounted at either end of the truss and a microphone located over the instrument panel provided the three input measurements. An accelerometer mounted on the instrument panel was used as the output measurement. The acceleration output $y(t)$ was measured in gravity units ($g = 9.8$ meters/sec^2) and all calculations were performed with a resolution bandwidth of $B_e = 5$ Hz and $n_d = 50$ averages.

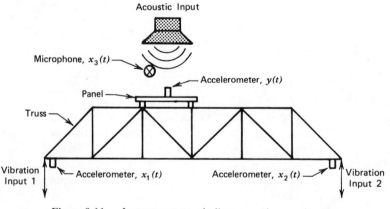

Figure 9.11   Instrument panel vibroacoustic experiment.

The autospectra of the three inputs and the output in the truss experiment are shown in Figure 9.12. Note that the response of the instrument panel is heavily concentrated in very narrow frequency bands corresponding to lightly damped vibration modes (resonances) of the instrument panel–truss assembly, as discussed in Section 1.3.3. Hence attention can be limited to these frequencies of resonance. The multiple coherence functions at the

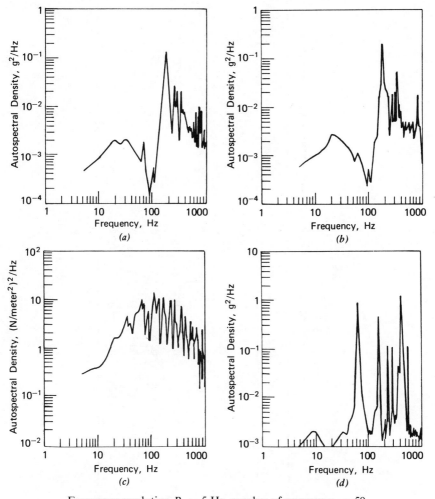

Frequency resolution $B_e = 5$ Hz, number of averages $n_d = 50$

Figure 9.12   Autospectra of instrument panel inputs and output. (*a*) Vibration input No. 1. (*b*) Vibration input No. 2. (*c*) Acoustic noise input. (*d*) Vibration output of panel. (These data are taken from Reference 9.2 with the permission of the Shock and Vibration Information Center and the author.)

Table 9.1

Partial Coherent Output Power Results for Instrument Panel Experiment

Frequency, Hz	Coherence Function Values				Outputs, g²/Hz		Partial Coherent Outputs, g²/Hz and (%)		
	$\gamma^2_{y:x}$	γ^2_{1y}	$\gamma^2_{2y\cdot 1}$	$\gamma^2_{3y\cdot 1,2}$	Total, G_{yy}	Coherent, $G_{y:x}$	$G_{y:1} = \gamma^2_{1y} G_{yy}$	$G_{y:2\cdot 1} = \gamma^2_{2y\cdot 1} G_{yy\cdot 1}$	$G_{y:3\cdot 1,2} = \gamma^2_{3y\cdot 1,2} G_{yy\cdot 1,2}$
68	0.81	0.78	0.09	0.04	0.91	0.74	0.71 (96%)	0.02 (3%)	0.01 (1%)
161	0.73	0.44	0.47	0.08	0.41	0.30	0.18 (60%)	0.11 (37%)	0.01 (3%)
230	0.72	0.15	0.56	0.26	0.14	0.10	0.02 (20%)	0.07 (70%)	0.01 (10%)
290	0.86	0.02	0.01	0.85	0.12	0.10	<0.01 (—)	<0.01 (—)	0.09 (90%)
420	0.64	0.05	0.06	0.60	1.2	0.77	0.06 (8%)	0.07 (9%)	0.64 (83%)

first five resonance frequencies are shown in Table 9.1. The multiple coherence function values are $\gamma_{y:x}^2 = 0.64$ to 0.86 which is acceptable; the lack of stronger multiple coherence values may be due to resolution bias errors at the resonance peaks, as discussed in Section 5.2.3. Also shown in Table 9.1 are the partial coherent output power values calculated in the following order: $x_1(t) =$ input from shaker 1 (left side of truss nearest instrument panel), $x_2(t) =$ input from shaker 2 (right side of truss), and $x_3(t) =$ input from acoustic noise.

From the results in Table 9.1, the partial coherent output due to $x_3(t)$, the acoustic noise input, is the easiest to interpret because $x_3(t)$ is reasonably uncorrelated with the vibration inputs $x_1(t)$ and $x_2(t)$. The results show that the statistically independent contribution of $x_3(t)$ to the coherent output $G_{y:x}$ increases from a negligible amount (roughly 1 %) at 68 Hz to a dominant contribution (83 %) at 420 Hz. This firmly establishes that the vibration inputs $x_1(t)$ and $x_2(t)$ cause most of the response $y(t)$ at the lower frequencies, and the acoustic input $x_3(t)$ is responsible for most of the response at the higher frequencies. The relative contributions of the two shaker inputs, however, must be interpreted with caution because $x_1(t)$ and $x_2(t)$ are highly correlated. Specifically, at 68 Hz, the results suggest that $x_1(t)$ is responsible for 96 % of the coherent output, but it is understood that much of this indicated input might be due to correlated contributions from $x_2(t)$. On the other hand, at 161 Hz, 60 % of the coherent output can be attributed to $x_1(t)$ and 37 % to the statistically independent contribution of $x_2(t)$ alone. This means that the total contribution of $x_2(t)$ is at least 37 %. At 230 Hz, the statistically independent contribution of $x_2(t)$ alone is 70 %, confirming that $x_2(t)$ dominates the response at this frequency independent of possible contributions associated with $x_1(t)$.

9.4.3 Summary of Errors

The random errors due to statistical sampling considerations in various partial coherence and partial coherent output quantities are presented in Chapter 11 and summarized in Table 9.2. Note that the errors in all cases are dependent on a coherence function as well as the number of averages n_d. Further note that the error expressions are useful only for relatively small errors ($\varepsilon_r < 0.20$). Error terms for higher order conditioned quantities are given by a methodical extension of the results in Table 9.2. The error formulas for the basic spectral calculations have been summarized previously in Table 3.2.

Table 9.2

Random Errors in Partial Coherence and Partial Coherent Output Spectrum Estimates

Function Being Estimated	Random Error, ε_r
$\hat{\gamma}_{xy}^2(f)$	$\dfrac{\sqrt{2}[1 - \hat{\gamma}_{xy}^2(f)]}{|\hat{\gamma}_{xy}(f)|\sqrt{n_d}}$
$\hat{G}_{y:x}(f)$	$\dfrac{[2 - \hat{\gamma}_{xy}^2(f)]^{1/2}}{|\hat{\gamma}_{xy}(f)|\sqrt{n_d}}$
$\hat{\gamma}_{x_2y\cdot x_1}^2(f)$	$\dfrac{\sqrt{2}[1 - \hat{\gamma}_{2y\cdot 1}^2(f)]}{|\hat{\gamma}_{2y\cdot 1}(f)|\sqrt{n_d - 1}}$
$\hat{G}_{y:x_2\cdot x_1}(f)$	$\dfrac{[2 - \hat{\gamma}_{2y\cdot 1}^2(f)]^{1/2}}{|\hat{\gamma}_{2y\cdot 1}(f)|\sqrt{n_d - 1}}$
$\hat{\gamma}_{x_3y\cdot x_1, x_2}^2(f)$	$\dfrac{\sqrt{2}[1 - \hat{\gamma}_{3y\cdot 1, 2}^2(f)]}{|\hat{\gamma}_{3y\cdot 1, 2}(f)|\sqrt{n_d - 2}}$
$\hat{G}_{y:x_3\cdot x_1, x_2}(f)$	$\dfrac{[2 - \hat{\gamma}_{3y\cdot 1, 2}^2(f)]^{1/2}}{|\hat{\gamma}_{3y\cdot 1, 2}(f)|\sqrt{n_d - 2}}$

Beyond the random errors in Table 9.2, there are numerous sources of potential bias errors that can either increase or decrease, on the average, the estimated coherence and coherent output power functions, as detailed in Sections 9.2 and 9.3. These bias errors as well as the random errors in Table 9.2 must be carefully considered in all applications.

REFERENCES

9.1 Halvorsen, W. G., and Bendat, J. S., "Noise Source Identification Using Coherent Output Power Spectra," *Sound and Vibration*, Vol. 9, p. 15, 1975.

9.2 Barrett, S., "On the Use of Coherence Functions to Evaluate Sources of Dynamic Excitation," *Shock and Vibration Bulletin*, No. 49, Part 1, p. 43, 1979.

CHAPTER 10

PROCEDURES FOR SOLVING MULTIPLE INPUT/OUTPUT PROBLEMS

Chapters 4 and 8 developed various useful relationships for single input/ output and multiple input/output problems. This chapter outlines iterative computational procedures that provide the basis for efficient digital computer analysis and simulation of multiple input/output data. Results are obtained for conditioned spectral quantities, optimum frequency response functions, decomposition of output spectra into physically meaningful quantities, partial coherence functions, and multiple coherence functions. As in Chapter 8, capital letters are used to denote finite Fourier transforms, and all developments are presented in terms of two-sided spectral density quantities.

10.1 FORMULATION OF MODELS

Relationships will be derived first for a two input system and will then be extended to a multiple input system with q inputs.

10.1.1 *Two Input Systems*

To lay the foundation for the general case to follow, consider the two input/ single output model of Figure 10.1 where the two inputs are arbitrary stationary (ergodic) random or transient random records. The two inputs are

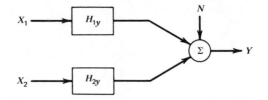

Figure 10.1   Two input/single output model for arbitrary inputs.

assumed to be correlated, but not perfectly correlated, so that $0 < \gamma_{12}^2 < 1$. For this model, from Chapter 8,

$$Y = H_{1y}X_1 + H_{2y}X_2 + N$$

$$H_{1y} = \frac{S_{1y \cdot 2}}{S_{11 \cdot 2}} \qquad H_{2y} = \frac{S_{2y \cdot 1}}{S_{22 \cdot 1}} \qquad (10.1)$$

$$S_{yy} = |H_{1y}|^2 S_{11} + H_{1y}^* H_{2y} S_{12} + H_{1y} H_{2y}^* S_{21} + |H_{2y}|^2 S_{22} + S_{nn}$$

Note that H_{1y} and H_{2y} are of equal complexity. Note also that S_{yy} contains five terms in all, where the terms involving S_{12} and S_{21} are difficult to interpret. With q inputs, the output spectrum would contain $q^2 + 1$ terms. It is not clear here how much of the spectral output of $y(t)$ is due to $x_1(t)$ alone, or the precise nature of the frequency response function to predict $y(t)$ from $x_1(t)$. It is also not clear how much of the spectral output of $y(t)$ is due to $x_{2 \cdot 1}(t)$ when the linear effects of $x_1(t)$ are removed from $x_2(t)$. These questions can be answered by considering an alternate model to Figure 10.1 using conditioned inputs.

Previous discussions in Chapter 8 using conditioned inputs that can be computed from the original given arbitrary inputs show that Figure 10.1 can be replaced by a new model which combines parts of Figures 8.4 and 8.5. This is illustrated in Figure 10.2 where the two inputs are uncorrelated. The output spectrum S_{yy} now contains three terms which are easy to interpret. With q inputs, the output spectrum would contain $q + 1$ terms.

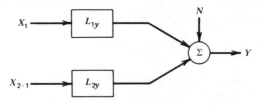

Figure 10.2   Two input/single output model for conditioned inputs.

From the developments in Section 8.3, the following equations apply to the model in Figure 10.2:

$$Y = L_{1y}X_1 + L_{2y}X_{2\cdot 1} + N$$

$$L_{1y} = \frac{S_{1y}}{S_{11}}$$

$$L_{2y} = \frac{S_{2y\cdot 1}}{S_{22\cdot 1}} \tag{10.2}$$

$$S_{yy} = |L_{1y}|^2 S_{11} + |L_{2y}|^2 S_{22\cdot 1} + S_{nn}$$

$$S_{nn} = S_{yy\cdot 1,2} = S_{yy}(1 - \gamma_{1y}^2)(1 - \gamma_{2y\cdot 1}^2)$$

Each of these quantities has a direct physical interpretation as follows:

$|L_{1y}|^2 S_{11} = \gamma_{1y}^2 S_{yy} =$ spectral output at $y(t)$ due to $x_1(t)$ passing through L_{1y} where L_{1y} is the optimum linear system to predict $y(t)$ from $x_1(t)$.

$|L_{2y}|^2 S_{22\cdot 1} = \gamma_{2y\cdot 1}^2 S_{yy\cdot 1} =$ spectral output at $y(t)$ due to $x_{2\cdot 1}(t)$ passing through L_{2y} where L_{2y} is the optimum linear system to predict $y(t)$ from $x_{2\cdot 1}(t)$.

$S_{nn} = S_{yy\cdot 1,2} =$ spectral output due to $n(t)$, representing all independent extraneous output noise at $y(t)$ that is not linearly due to $x_1(t)$ or $x_2(t)$ passing through optimum systems to predict $y(t)$.

Note that the terms S_{nn} and S_{yy} are the same quantities in Figures 10.1 and 10.2.

By interchanging the records, an alternate model to Figure 10.2 is obtained, as shown in Figure 10.3. The output spectrum is now

$$S_{yy} = \left|\frac{S_{2y}}{S_{22}}\right|^2 S_{22} + \left|\frac{S_{1y\cdot 2}}{S_{11\cdot 2}}\right|^2 S_{11\cdot 2} + S_{nn}$$

$$= \gamma_{2y}^2 S_{yy} + \gamma_{1y\cdot 2}^2 S_{yy\cdot 2} + S_{nn} \tag{10.3}$$

where

$\gamma_{2y}^2 S_{yy} =$ spectral output at $y(t)$ due to $x_2(t)$ passing through the optimum linear system (S_{2y}/S_{22}) to predict $y(t)$ from $x_2(t)$.

$\gamma_{1y\cdot 2}^2 S_{yy\cdot 2} =$ spectral output at $y(t)$ due to $x_{1\cdot 2}(t)$ passing through the optimum linear system $(S_{1y\cdot 2}/S_{11\cdot 2})$ to predict $y(t)$ from $x_{1\cdot 2}(t)$.

No change occurs in S_{nn} and S_{yy}.

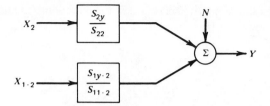

Figure 10.3   Alternate two input/single output model for conditioned inputs.

These results show that it is important to select the order of the input records in a definite way since it determines how the results are interpreted. If either order is physically possible, it is recommended that the ordinary coherence functions be computed between $x_1(t)$ and $y(t)$, and between $x_2(t)$ and $y(t)$, and that $x_1(t)$ be selected as that record where $\gamma_{1y}^2 \geq \gamma_{2y}^2$. Then $x_{2 \cdot 1}(t)$ will remove the linear effects of the stronger $x_1(t)$ from $x_2(t)$ so as to better allow the prediction of $y(t)$ from $x_{2 \cdot 1}(t)$, as shown in Figure 10.2. Of course, if it is known that $x_1(t)$ should follow $x_2(t)$, then $x_2(t)$ should be chosen as the first record and $x_{1 \cdot 2}(t)$ becomes the second record, as shown in Figure 10.3.

An *incorrect* model that might be proposed to replace Figure 10.1 is shown in Figure 10.4. For this model, the two conditioned inputs can be computed from the original given data. However, the noise term M will be different than the previous N. In Figure 10.4,

$$Y = \left(\frac{S_{1y \cdot 2}}{S_{11 \cdot 2}}\right) X_{1 \cdot 2} + \left(\frac{S_{2y \cdot 1}}{S_{22 \cdot 1}}\right) X_{2 \cdot 1} + M$$

$$S_{yy} = \gamma_{1y \cdot 2}^2 S_{yy \cdot 2} + \gamma_{2y \cdot 1}^2 S_{yy \cdot 1} + S_{mm}$$
$$= \gamma_{1y \cdot 2}^2 (1 - \gamma_{2y}^2) S_{yy} + \gamma_{2y \cdot 1}^2 (1 - \gamma_{1y}^2) S_{yy} + S_{mm}$$

Now

$$S_{mm} = S_{yy}[1 - \gamma_{1y \cdot 2}^2 (1 - \gamma_{2y}^2) - \gamma_{2y \cdot 1}^2 (1 - \gamma_{1y}^2)]$$

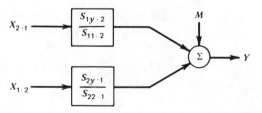

Figure 10.4   Incorrect two input/single output model.

But the general value for S_{nn} is known from Equation 8.69 to be

$$S_{nn} = S_{yy}(1 - \gamma_{1y}^2)(1 - \gamma_{2y \cdot 1}^2) \neq S_{mm}$$

Thus Figure 10.4 is *not* suitable to replace Figure 10.1.

10.1.2 *Multiple Input Systems*

The general multiple input/output model is illustrated in Figure 10.5, where the terms $\{X_i\}$, $i = 1, 2, \ldots, q$, represent finite Fourier transforms of the input records selected in a particular order. The finite Fourier transform of the output record is represented by $Y = X_{q+1}$. Constant parameter linear frequency response functions are denoted by $\{H_{iy}\}$, $i = 1, 2, \ldots, q$, where the input index *precedes* the output index. All possible deviations from the ideal overall model are accounted for in the finite Fourier transform N of the unknown extraneous output noise. Similar models can be formulated by interchanging the order of the inputs or by selecting different output records.

In practice, no restrictions are placed on the various autospectral and cross-spectral density functions between the given records other than they exist and can be estimated from the available data. None of the individual linear outputs from any X_i passing through H_{iy} is assumed to be directly measurable, nor is the output noise term N directly measurable. In general, all of the $\{H_{iy}\}$ are assumed unknown and are to be estimated. Of course,

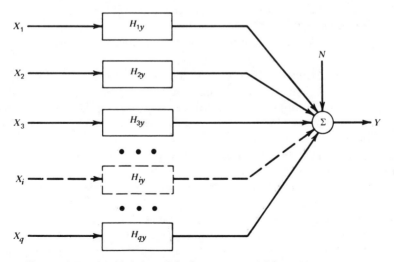

Figure 10.5 Multiple input/single output model for arbitrary inputs.

in a particular problem, some of the H_{iy} might be known or measurable. These linear frequency response functions $\{H_{iy}\}$ do not necessarily represent the actual physically realizable characteristics of a given situation. They can be merely mathematical data processing results that are used to relate Y to the various $\{X_i\}$ by optimum linear least-squares operations.

Four conditions are required for this model to be well defined:

1. None of the ordinary coherence functions between any pair of input records should equal unity. If this occurs, then the two inputs contain redundant information and one of the inputs should be eliminated from the model. This consideration allows distributed input systems to be studied as discrete inputs.
2. None of the ordinary coherence functions between any input and the total output should equal unity. If this occurs, then the other inputs are not contributing to this output and the model should be considered simply as a single input/single output model.
3. The multiple coherence function between any input and the other inputs, excluding the given input, should not equal unity. If this occurs, then the input in question can be obtained by linear operations from the other inputs, that is, it is not providing any new information to the output and should be eliminated from the model.
4. The multiple coherence function between the output and the given inputs in a practical situation should be sufficiently high, say above 0.5, in order for the theoretical assumptions and later conclusions to be reasonable. Otherwise, some inputs are probably being omitted or nonlinear effects are large. This value of 0.5 is not precise, but it is a matter of engineering and statistical judgment based on the physical environment and the amount of data available for analysis.

An alternate model to Figure 10.5 occurs by replacing the original given input records by an ordered set of conditioned input records. These are denoted by the finite Fourier transforms $\{X_{i\cdot(i-1)!}\}$, $i = 1, 2, \ldots, q$, selected in the order shown in Figure 10.6. For any i, the term $X_{i\cdot(i-1)!}$ represents X_i conditioned on the previous $X_1, X_2, \ldots, X_{i-1}$, that is, when the linear effects of X_1 to X_{i-1} have been removed from X_i by optimum linear least-squares prediction techniques. These ordered conditioned input records will be mutually uncorrelated, a property not generally satisfied by the original given input records.

The set of optimum constant parameter linear frequency response functions in Figure 10.6 is denoted by $\{L_{iy}\}$, $i = 1, 2, \ldots, q$. These $\{L_{iy}\}$ are generally quite different than the $\{H_{iy}\}$ in Figure 10.5. The output terms N and Y are exactly the same in both figures. Note that $q!$ different models are

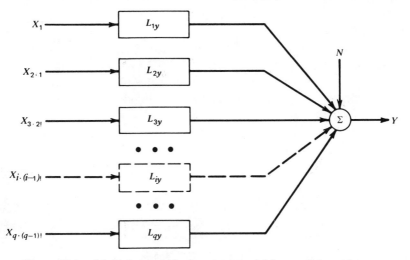

Figure 10.6   Multiple input/single output model for conditioned inputs.

possible depending on which record is chosen as X_1, which is chosen as $X_{2 \cdot 1}$, then $X_{3 \cdot 2!}$, and so on. The analysis to follow is based on choosing a particular order of the input records and remaining with this order. Similar results apply to any other desired ordering.

10.1.3   Fourier Transforms and Spectral Functions

Reviewing from Chapter 3, for any pair of records $x(t)$ and $y(t)$ of length T, their finite Fourier transforms will be defined by

$$X(f, T) = \int_0^T x(t)e^{-j2\pi ft}\, dt$$

$$Y(f, T) = \int_0^T y(t)e^{-j2\pi ft}\, dt$$

(10.4)

When computations are carried out digitally, these formulas give discrete values of f spaced $\Delta f = (1/T)$ apart. It is assumed here that $x(t)$ and $y(t)$ are members of stationary (ergodic) random or transient random processes such that expected values can be taken over the members. As in Equation 8.8, this gives the two-sided autospectral density functions

$$S_{xx}(f) = \frac{1}{T} E[X^*(f, T)X(f, T)] \qquad S_{yy}(f) = \frac{1}{T} E[Y^*(f, T)Y(f, T)]$$

(10.5a)

and the cross-spectral density function

$$S_{xy}(f) = \frac{1}{T} E[X^*(f, T)Y(f, T)] \tag{10.5b}$$

In practice, of course, one obtains only estimates of these quantities since the expected value operation is taken over a finite number of sample records. Note also that for transient random data, the factor $(1/T)$ would be omitted. From Section 3.4.2, an estimate of $\hat{S}_{xy}(f)$ is usually obtained by computing

$$\hat{S}_{xy}(f) = \frac{1}{n_d T} \sum_{k=1}^{n_d} X_k^*(f, T)Y_k(f, T) \tag{10.5c}$$

where n_d is the number of independent (disjoint) sample records each of length T. The total record length T_{total} is effectively $T_{total} = n_d T$. When T_{total} is fixed, compromise choices must be made between values of T and n_d. As will be shown in Chapter 11, to reduce bias errors, T should be as large as possible, whereas to reduce random errors, n_d should be as large as possible.

In the following material, it will be assumed that proper choices have been made of desired parameters to obtain good estimates of the various auto-spectra and cross-spectra between all of the $(q + 1)$ records representing the original given q input processes and the output process. The expected value (averaging) operation required by Equations 10.5a, b, and c is performed using sample records representing these $(q + 1)$ processes only. Later computations of conditioned spectral quantities do not require any further choices of parameters or any further expected value operations, although these quantities will be defined using expected values. Instead, the various conditioned spectral quantities can be computed by iterative algebraic formulas to be derived in this chapter.

Finite Fourier transforms (F.T.) of original records and of conditioned records over a long record length T will be denoted by capital letters as follows, where the dependence on T and f will be omitted for simplicity in notation:

$$X_i = \text{F.T. of } x_i(t) \qquad \text{for } i = 1, 2, \ldots, q + 1$$
$$Y = X_{q+1} = \text{F.T. of } y(t) = x_{q+1}(t) \tag{10.6}$$

For any $r = 1, 2, \ldots, q$, where $i > r$, let

$$X_{i \cdot r!} = X_{i \cdot 1, 2, \ldots, r} = \text{F.T. of } x_i(t) \text{ conditioned on}$$
$$\text{the previous } x_1(t) \text{ to } x_r(t) \tag{10.7}$$
$$N = X_{q+1 \cdot q!} = Y_{y \cdot q!} = \text{F.T. of } n(t)$$

For any i, the quantity $X_{i \cdot r!} = 0$ if $r \geq i$ since the effects of $x_i(t)$ will have been removed from $x_i(t)$ if $r \geq i$.

For the original records, their autospectral and cross-spectral density functions from Equations 10.5a and b are

$$(T)S_{ii} = E[X_i^* X_i] = E[|X_i|^2]$$
$$(T)S_{ij} = E[X_i^* X_j] \text{ when } i \neq j$$

(10.8)

$$(T)S_{yy} = E[|Y|^2] = E[|X_{q+1}|^2]$$
$$(T)S_{iy} = E[X_i^* Y] = E[X_i^* X_{q+1}]$$

(10.9)

For conditioned records, their conditioned autospectral density functions are defined by

$$(T)S_{ii \cdot r!} = E[|X_{i \cdot r!}|^2]$$
$$(T)S_{nn} = (T)S_{(q+1)(q+1) \cdot q!} = E[|N|^2]$$

(10.10)

The conditioned cross-spectral density functions are defined by

$$(T)S_{ij \cdot r!} = E[X_{i \cdot r!}^* X_{j \cdot r!}]$$

(10.11)

when $i \neq j$ with both $i > r$ and $j > r$. Note that

$$S_{ij \cdot r!}^* = S_{ji \cdot r!}$$
$$S_{iy \cdot (i-1)!} = S_{i(q+1) \cdot (i-1)!}$$

(10.12)

For all i and j, the quantities $S_{ij \cdot r!} = 0$ if either $r \geq i$ or $r \geq j$.

In Figure 10.6, if all q ordered conditioned inputs are used, then the output $Y = X_{q+1}$ can be expressed as

$$X_{q+1} = \sum_{i=1}^{q} L_{iy} X_{i \cdot (i-1)!} + X_{(q+1) \cdot q!}$$

(10.13)

Suppose that Figure 10.6 is modified by considering only the effect of the first r ordered conditioned inputs where $r \leq q$, but still leaving $x_{q+1}(t)$ as the output. Then the noise term becomes $X_{(q+1) \cdot r!}$ and Equation 10.13 takes the form

$$X_{q+1} = \sum_{i=1}^{r} L_{iy} X_{i \cdot (i-1)!} + X_{(q+1) \cdot r!}$$

(10.14)

Substitution of X_j for X_{q+1} gives the more general equation

$$X_j = \sum_{i=1}^{r} L_{ij} X_{i \cdot (i-1)!} + X_{j \cdot r!}$$

(10.15)

It follows from Equation 10.15 that for any $j \geq i$ and any $r < i$

$$E[X_{i \cdot r!}^* X_j] = E[X_{i \cdot r!}^* X_{j \cdot r!}]$$

(10.16)

since all other expected values will be zero. Hence Equations 10.10 and 10.11 can be computed by

$$
\begin{aligned}
(T)S_{ii \cdot r!} &= E[X^*_{i \cdot r!} X_i] \\
(T)S_{ij \cdot r!} &= E[X^*_{i \cdot r!} X_j]
\end{aligned}
\tag{10.17}
$$

This also proves that

$$
\begin{aligned}
(T)S_{yy \cdot r!} &= E[X^*_{(q+1) \cdot r!} X_{q+1}] \\
(T)S_{iy \cdot r!} &= E[X^*_{i \cdot r!} X_{q+1}]
\end{aligned}
\tag{10.18}
$$

Equations 10.17 and 10.18 are important theoretical formulas which are simpler to apply than the original definitions. In all of the above formulas, the factor T should be omitted when quantities are in "energy" units instead of in "power" units, as appropriate for transient random data versus stationary random data.

10.2 OPTIMUM SYSTEM RELATIONSHIPS

Of concern now are the procedures for calculating the optimum systems $\{L_{iy}\}$, $i = 1, 2, \ldots, q$, which relate the conditioned input records to the output in Figure 10.6. Also of concern are the relationships between these $\{L_{iy}\}$ systems and the $\{H_{iy}\}$ systems relating the original input records to the output in Figure 10.5. These matters will now be detailed.

10.2.1 Optimum Systems for Conditioned Inputs

Referring back to Equation 10.13, this result may be written as

$$
Y = \sum_{i=1}^{q} L_{iy} X_{i \cdot (i-1)!} + N
\tag{10.19}
$$

where it is convenient here to replace X_{q+1} by Y and $X_{(q+1) \cdot q!}$ by N. Let N_{iy} represent the noise quantity

$$
N_{iy} = Y - L_{iy} X_{i \cdot (i-1)!}
\tag{10.20}
$$

which is the difference between the output Y and the output due to the conditioned input $X_{i \cdot (i-1)!}$ passing through L_{iy}. Then

$$
\begin{aligned}
|N_{iy}|^2 = |Y|^2 &- L^*_{iy}[X^*_{i \cdot (i-1)!} Y] \\
&- L_{iy}[Y^* X_{i \cdot (i-1)!}] + L^*_{iy} L_{iy}[|X_{i \cdot (i-1)!}|^2]
\end{aligned}
\tag{10.21}
$$

Taking expected values of both sides and dividing by T yields

$$S_{n_i n_i} = S_{yy} - L_{iy}^* S_{iy \cdot (i-1)!}$$
$$- L_{iy} S_{iy \cdot (i-1)!}^* + L_{iy}^* L_{iy} S_{ii \cdot (i-1)!} \qquad (10.22)$$

Equation 10.22 reflects the mean square error for any L_{iy}. The optimum system, defined as that linear system which produces minimum mean square system error, can be obtained by setting the partial derivative of Equation 10.22, with respect to L_{iy}, equal to zero, holding L_{iy}^* fixed. This proves, as in Section 5.1.1, that

$$L_{iy} = \frac{S_{iy \cdot (i-1)!}}{S_{ii \cdot (i-1)!}} \qquad i = 1, 2, \ldots, q \qquad (10.23)$$

If $Y = X_{q+1}$ is replaced by X_j, then L_{iy} becomes L_{ij}, where

$$L_{ij} = \frac{S_{ij \cdot (i-1)!}}{S_{ii \cdot (i-1)!}} \qquad i, j = 1, 2, \ldots, q, q+1 \qquad (10.24)$$

Note that

$$\begin{aligned} L_{ii} &= 1 \qquad \text{for any } i \\ L_{ij} &= 0 \qquad \text{for } j < i \\ L_{ij}^* &= L_{ji} \end{aligned} \qquad (10.25)$$

Special cases of Equation 10.24 are

$$i = 1: \quad L_{1j} = \frac{S_{1j}}{S_{11}} \qquad j = 1, 2, \ldots, q+1$$

$$i = 2: \quad L_{2j} = \frac{S_{2j \cdot 1}}{S_{22 \cdot 1}} \qquad j = 2, 3, \ldots, q+1 \qquad (10.26)$$

$$i = 3: \quad L_{3j} = \frac{S_{3j \cdot 2!}}{S_{33 \cdot 2!}} \qquad j = 3, 4, \ldots, q+1$$

The last term where $i = q$ and $j = q + 1$ gives

$$L_{q(q+1)} = L_{qy} = \frac{S_{qy \cdot (q-1)!}}{S_{qq \cdot (q-1)!}} \qquad (10.27)$$

It follows also from Equations 10.19 and 10.23 that the noise term will be uncorrelated with the ordered conditioned input terms. Equations 10.23 and 10.24 show that it is essential in this work to be able to compute the conditioned cross-spectral density functions $S_{ij \cdot (i-1)!}$ for any i and j. Recommended ways to compute these functions by special iterative computational algorithms are developed in Section 10.3.2.

10.2.2 Optimum Systems for Original Inputs

Unlike the set of ordered conditioned inputs in Figure 10.6, the set of finite Fourier transforms given by $\{X_i\}$, $i = 1, 2, \ldots, q$, in Figure 10.5 do not, in general, represent mutually uncorrelated records. The output record $Y = X_{q+1}$ is a linear sum of such terms, namely

$$Y = \sum_{i=1}^{q} H_{iy} X_i + N \tag{10.28}$$

which by changing the index of summation from i to j becomes

$$Y = \sum_{j=1}^{q} H_{jy} X_j + N \tag{10.29}$$

Now multiply Equation 10.29 in turn by X_i^*, $X_{i\cdot 1}^*$, $X_{i\cdot 2!}^*$, and so on, and take expected values of both sides. From Equation 10.17, this yields the results

$$S_{iy} = \sum_{j=1}^{q} H_{jy} S_{ij} \qquad i = 1, 2, \ldots, q \tag{10.30}$$

$$S_{iy\cdot 1} = \sum_{j=2}^{q} H_{jy} S_{ij\cdot 1} \qquad i = 2, 3, \ldots, q$$

$$S_{iy\cdot 2!} = \sum_{j=3}^{q} H_{jy} S_{ij\cdot 2!} \qquad i = 3, 4, \ldots, q \tag{10.31}$$

which continues up to the last term where

$$S_{qy\cdot(q-1)!} = H_{qy} S_{qq\cdot(q-1)!} \tag{10.32}$$

Thus

$$H_{qy} = \frac{S_{qy\cdot(q-1)!}}{S_{qq\cdot(q-1)!}} \tag{10.33}$$

which is the same as L_{qy} in Equation 10.23. The optimum system H_{iy} for any i follows by merely interchanging $x_i(t)$ with $x_q(t)$. Hence

$$H_{iy} = \frac{S_{iy\cdot 1, 2, \ldots, (i-1), (i+1), \ldots, q}}{S_{ii\cdot 1, 2, \ldots, (i-1), (i+1), \ldots, q}} \qquad i = 1, 2, \ldots, q \tag{10.34}$$

Note that the $\{H_{iy}\}$ of Equation 10.34 will contain many more terms than the $\{L_{iy}\}$ of Equation 10.23 except for $i = q$ where $H_{qy} = L_{qy}$ so that it is computationally more efficient to compute the $\{L_{iy}\}$.

10.2.3 *Relations Between Optimum Systems*

Relations between these two sets of optimum systems will now be derived. Multiply Equation 10.29 by $X^*_{i \cdot (r-1)!}$, where $r \le i \le q$, and take expected values of both sides. This proves that the quantity $S_{iy \cdot (r-1)!}$ satisfies the general expression

$$S_{iy \cdot (r-1)!} = \sum_{j=r}^{q} H_{jy} S_{ij \cdot (r-1)!} \qquad (10.35)$$

For $r = i$, Equation 10.35 becomes

$$S_{iy \cdot (i-1)!} = \sum_{j=i}^{q} H_{jy} S_{ij \cdot (i-1)!} \qquad (10.36)$$

Divide by $S_{ii \cdot (i-1)!}$ and substitute from Equations 10.23 and 10.24 to obtain

$$L_{iy} = \sum_{j=i}^{q} L_{ij} H_{jy} \qquad i = 1, 2, \ldots, q \qquad (10.37)$$

Equation 10.37 shows exactly how to determine $\{L_{iy}\}$ from $\{H_{iy}\}$. Conversely, a convenient way to determine $\{H_{iy}\}$ from $\{L_{iy}\}$ is to work backward as follows:

$$H_{qy} = L_{qy}$$
$$H_{iy} = L_{iy} - \sum_{j=i+1}^{q} L_{ij} H_{jy} \qquad i = (q-1), (q-2), \ldots, 1 \qquad (10.38)$$

For example, if $q = 4$, Equation 10.37 gives

$$\begin{aligned}
L_{1y} &= H_{1y} + L_{12} H_{2y} + L_{13} H_{3y} + L_{14} H_{4y} \\
L_{2y} &= \phantom{H_{1y} + {}} H_{2y} + L_{23} H_{3y} + L_{24} H_{4y} \\
L_{3y} &= \phantom{H_{1y} + L_{12} H_{2y} + {}} H_{3y} + L_{34} H_{4y} \\
L_{4y} &= \phantom{H_{1y} + L_{12} H_{2y} + L_{13} H_{3y} + {}} H_{4y}
\end{aligned} \qquad (10.39)$$

On the other hand, Equation 10.38 gives

$$\begin{aligned}
H_{4y} &= L_{4y} \\
H_{3y} &= L_{3y} - L_{34} H_{4y} \\
H_{2y} &= L_{2y} - L_{23} H_{3y} - L_{24} H_{4y} \\
H_{1y} &= L_{1y} - L_{12} H_{2y} - L_{13} H_{3y} - L_{14} H_{4y}
\end{aligned} \qquad (10.40)$$

where previous results should be substituted in succeeding equations. This example is illustrated in Figures 10.7 and 10.8.

Results in Equation 10.39 show exactly how the conditioned inputs X_1, $X_{2 \cdot 1}$, $X_{3 \cdot 2!}$, and $X_{4 \cdot 3!}$, taken in that order, pass through H_{1y}, H_{2y},

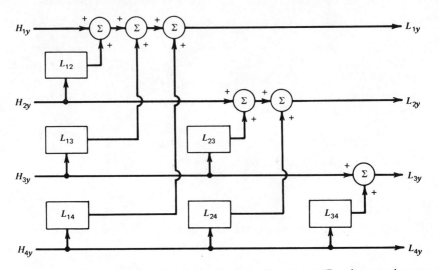

Figure 10.7 Determination of optimum L_{iy} from H_{iy} systems. (Results extend to any number of systems.)

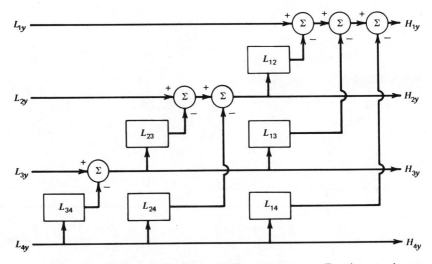

Figure 10.8 Determination of optimum H_{iy} from L_{iy} systems. (Results extend to any number of systems.)

H_{3y}, and H_{4y} as a function of frequency. For example, $X_{2 \cdot 1}$ does not go through H_{3y} if the term $L_{23} = 0$, and it does not go through H_{4y} if $L_{24} = 0$. Similarly, $X_{3 \cdot 2!}$ does not go through H_{4y} if $L_{34} = 0$.

10.3 COMPUTATIONAL ALGORITHMS

The actual computations required to evaluate multiple input/output problems by the procedures developed in this chapter reduce to a straightforward set of iterative operations. These operations are generally easier and faster to execute than the traditional matrix solutions to multiple input problems. Furthermore, they give insight into the physical meaning of the results, as will now be discussed. Relevant previous publications are References 10.1, 10.2, and 10.3.

10.3.1 *Conditioned Records from Original Records*

Return to Equation 10.15 where

$$X_j = \sum_{i=1}^{r} L_{ij} X_{i \cdot (i-1)!} + X_{j \cdot r!} \tag{10.41}$$

If r is replaced by $(r - 1)$, then Equation 10.41 becomes

$$X_j = \sum_{i=1}^{r-1} L_{ij} X_{i \cdot (i-1)!} + X_{j \cdot (r-1)!} \tag{10.42}$$

Subtraction yields the algorithm

$$X_{j \cdot r!} = X_{j \cdot (r-1)!} - L_{rj} X_{r \cdot (r-1)!} \tag{10.43}$$

Special cases of Equation 10.43 are

$$
\begin{array}{lll}
r = 1: & X_{j \cdot 1} = X_j - L_{1j} X_1 & j = 2, 3, \ldots, q + 1 \\
r = 2: & X_{j \cdot 2!} = X_{j \cdot 1} - L_{2j} X_{2 \cdot 1} & j = 3, 4, \ldots, q + 1 \quad (10.44) \\
r = 3: & X_{j \cdot 3!} = X_{j \cdot 2!} - L_{3j} X_{3 \cdot 2!} & j = 4, 5, \ldots, q + 1
\end{array}
$$

The last term where $r = q$ and $j = q + 1$ is

$$X_{(q+1) \cdot q!} = X_{(q+1) \cdot (q-1)!} - L_{q(q+1)} X_{q \cdot (q-1)!} \tag{10.45}$$

Observe that this is an iterative procedure to compute Fourier transforms of ordered conditioned records. This procedure is illustrated in Figure 10.9

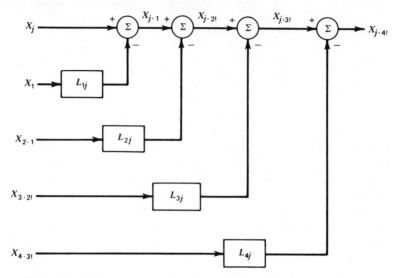

Figure 10.9 Algorithm to compute Fourier transforms of ordered conditioned inputs. (Results extend to any number of inputs.)

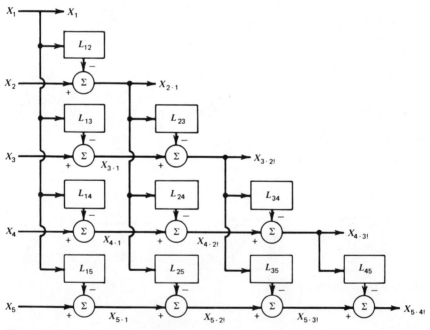

Figure 10.10 Determination of ordered conditioned records from original inputs. (Results extend to any number of inputs.)

250

and can be extended to any number of records. Various index choices in Equation 10.44 give

$$X_{2 \cdot 1} = X_2 - L_{12} X_1$$
$$X_{3 \cdot 2!} = X_{3 \cdot 1} - L_{23} X_{2 \cdot 1}$$
$$= X_3 - L_{13} X_1 - L_{23} X_{2 \cdot 1}$$
$$X_{4 \cdot 3!} = X_{4 \cdot 2!} - L_{34} X_{3 \cdot 2!}$$
$$= X_{4 \cdot 1} - L_{24} X_{2 \cdot 1} - L_{34} X_{3 \cdot 2!}$$
$$= X_4 - L_{14} X_1 - L_{24} X_{2 \cdot 1} - L_{34} X_{3 \cdot 2!}$$

$$(10.46)$$

and so on to as many terms as required. These results are the computational algorithms needed to obtain the ordered conditioned input records in Figure 10.6 from the original arbitrary input records in Figure 10.5. Figure 10.10 shows schematically the special case of five records and is a procedure that can be extended to any number of records.

The bottom line in Figure 10.10 can be solved for X_5 rather than for $X_{5 \cdot 4!}$ by using Equation 10.41 instead of Equation 10.43. In particular, the positions of X_5 and $X_{5 \cdot 4!}$ can be interchanged by summing, instead of subtracting, the terms on the bottom line, leaving everything else alone. Considered as a four input/output problem, X_5 can be replaced by Y and $X_{5 \cdot 4!}$ by N. Here

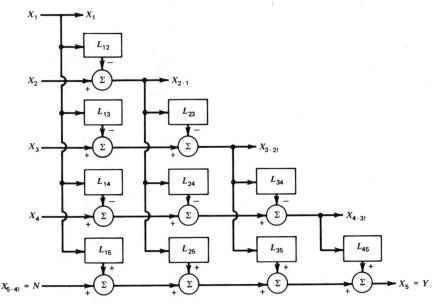

Figure 10.11 Fourier transform relationships for multiple input/single output model. (Results extend to any number of inputs.)

the systems $\{L_{i5}\}$ for $i = 1, 2, 3, 4$ are the optimum $\{L_{iy}\}$. These points have been noted in Figure 10.11 and show schematically how to relate four input/output records. Similar procedures apply to any number of records.

10.3.2 Conditioned Spectral Density Functions

To compute the conditioned spectral density functions, multiply Equation 10.43 by X_i^*, take the expected value of both sides, and divide by T. This gives the general algorithm

$$S_{ij\cdot r!} = S_{ij\cdot(r-1)!} - L_{rj}S_{ir\cdot(r-1)!} \tag{10.47}$$

where L_{rj} is calculated from preceding terms by Equation 10.51. Special cases are

$$
\begin{array}{lll}
r = 1: & S_{ij\cdot 1} = S_{ij} - L_{1j}S_{i1} & i, j = 2, 3, \ldots, q + 1 \\
r = 2: & S_{ij\cdot 2!} = S_{ij\cdot 1} - L_{2j}S_{i2\cdot 1} & i, j = 3, 4, \ldots, q + 1 \\
r = 3: & S_{ij\cdot 3!} = S_{ij\cdot 2!} - L_{3j}S_{i3\cdot 2!} & i, j = 4, 5, \ldots, q + 1
\end{array} \tag{10.48}
$$

The last term where $r = q$ and $i = j = (q + 1)$ is

$$S_{(q+1)(q+1)\cdot q!} = S_{(q+1)(q+1)\cdot(q-1)!} - L_{q(q+1)}S_{(q+1)q\cdot(q-1)!} \tag{10.49}$$

The iterative nature of Equation 10.47 to compute conditioned spectral density functions is pictured in Figure 10.12. This algorithm is applied in Reference 10.4.

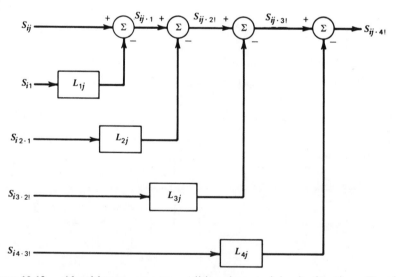

Figure 10.12 Algorithm to compute conditioned spectral density functions. (Results extend to any number of inputs.)

Consider Equation 10.47 for the terms where $i = j$, namely

$$S_{jj \cdot r!} = S_{jj \cdot (r-1)!} - L_{rj} S_{jr \cdot (r-1)!} \tag{10.50}$$

From Equation 10.24 with i replaced by r,

$$L_{rj} = \frac{S_{rj \cdot (r-1)!}}{S_{rr \cdot (r-1)!}} \tag{10.51}$$

Hence

$$L_{rj}^* = \frac{S_{jr \cdot (r-1)!}}{S_{rr \cdot (r-1)!}} \tag{10.52}$$

Substitution into Equation 10.50 now yields

$$S_{jj \cdot r!} = S_{jj \cdot (r-1)!} - |L_{rj}|^2 S_{rr \cdot (r-1)!} \tag{10.53}$$

representing the special iterative form taken by Equation 10.47 when the terms are autospectral density functions. This procedure is pictured in Figure 10.13. Special cases of Equation 10.53 are of interest:

$$
\begin{array}{llll}
r = 1: & S_{jj \cdot 1} = S_{jj} - |L_{1j}|^2 S_{11} & j = 2, 3, \ldots, q + 1 & \\
r = 2: & S_{jj \cdot 2!} = S_{jj \cdot 1} - |L_{2j}|^2 S_{22 \cdot 1} & j = 3, 4, \ldots, q + 1 & (10.54) \\
r = 3: & S_{jj \cdot 3!} = S_{jj \cdot 2!} - |L_{3j}|^2 S_{33 \cdot 2!} & j = 4, 5, \ldots, q + 1 &
\end{array}
$$

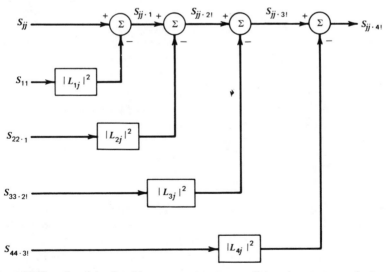

Figure 10.13 Special algorithm to compute conditioned autospectral density functions. (Results extend to any number of inputs.)

and so on. It follows that

$$j = 2: \quad S_{22 \cdot 1} = S_{22} - |L_{12}|^2 S_{11}$$

$$j = 3: \quad S_{33 \cdot 2!} = S_{33 \cdot 1} - |L_{23}|^2 S_{22 \cdot 1} \qquad (10.55)$$

$$j = 4: \quad S_{44 \cdot 3!} = S_{44 \cdot 2!} - |L_{34}|^2 S_{33 \cdot 2!}$$

and so on. These last results can also be derived from Equation 10.46. Thus the above formulas with their special cases yield the optimum systems of Equations 10.23 and 10.24 to solve multiple input/output problems. Figure 10.14 shows schematically the special case of five records. Note that this procedure extends to any number of records in exactly the same way as Figure 10.10.

As a further result of these formulas, Equation 10.53 can be traced back to prove that for any $r < j$ where $r \le q$

$$S_{jj} = \sum_{i=1}^{r} |L_{ij}|^2 S_{ii \cdot (i-1)!} + S_{jj \cdot r!} \qquad (10.56)$$

When $r = q$ and $j = q + 1$, this becomes

$$S_{(q+1)(q+1)} = \sum_{i=1}^{q} |L_{i(q+1)}|^2 S_{ii \cdot (i-1)!} + S_{(q+1)(q+1) \cdot q!} \qquad (10.57)$$

which is the same as

$$S_{yy} = \sum_{i=1}^{q} |L_{iy}|^2 S_{ii \cdot (i-1)!} + S_{nn} \qquad (10.58)$$

These equations decompose autospectral density functions into an ordered set of uncorrelated components. For example, if $q = 4$, one obtains

$$S_{55} = \sum_{i=1}^{4} |L_{i5}|^2 S_{ii \cdot (i-1)!} + S_{55 \cdot 4!} \qquad (10.59)$$

where $S_{55} = S_{yy}$ and $S_{55 \cdot 4!} = S_{nn}$. Thus S_{55} is decomposed into the sum of five terms as pictured in the four input/output model of Figure 10.15. This procedure can be extended to any number of records. Figure 10.15 is equivalent to the combination of four separate single input/output models as shown in Figure 10.16, where every term can be interpreted physically.

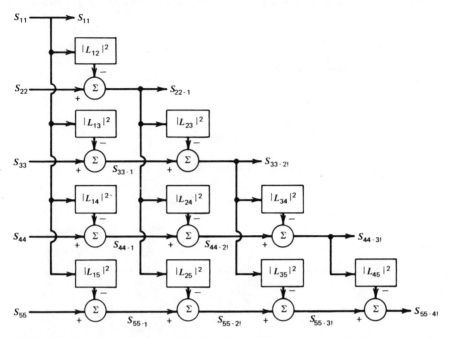

Figure 10.14 Determination of ordered conditioned autospectra from original autospectra. (Results extend to any number of inputs.)

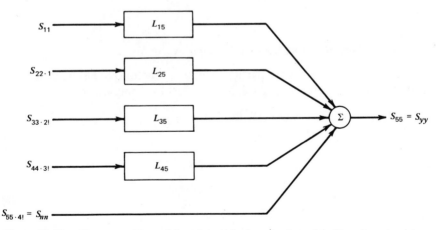

Figure 10.15 Decomposition of four input/single output model. (Results extend to any number of inputs.)

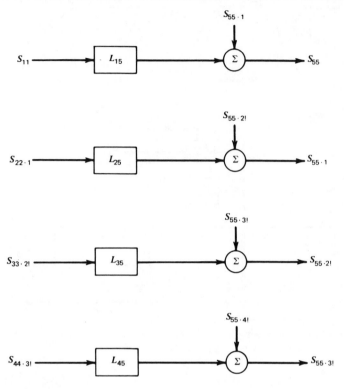

Figure 10.16 Single input/output models equivalent to Figure 10.15.

10.3.3 *Partial Coherence Functions and Noise Spectra*

Definitions of successive partial coherence functions are as follows, where the ordinary coherence function is given first for comparison:

$$\gamma_{iy}^2 = \frac{|S_{iy}|^2}{S_{ii}S_{yy}} \qquad i = 1, 2, \ldots, q$$

$$\gamma_{iy\cdot 1}^2 = \frac{|S_{iy\cdot 1}|^2}{S_{ii\cdot 1}S_{yy\cdot 1}} \qquad i = 2, 3, \ldots, q$$

$$\gamma_{iy\cdot 2!}^2 = \frac{|S_{iy\cdot 2!}|^2}{S_{ii\cdot 2!}S_{yy\cdot 2!}} \qquad i = 3, 4, \ldots, q \qquad (10.60)$$

$$\gamma_{iy\cdot 3!}^2 = \frac{|S_{iy\cdot 3!}|^2}{S_{ii\cdot 3!}S_{yy\cdot 3!}} \qquad i = 4, 5, \ldots, q$$

and so on. In particular, for $i = 1, 2, \ldots, q$,

$$\gamma^2_{iy\cdot(i-1)!} = \frac{|S_{iy\cdot(i-1)!}|^2}{S_{ii\cdot(i-1)!}S_{yy\cdot(i-1)!}} \tag{10.61}$$

From Equation 10.23 it follows that

$$|L_{iy}|^2 S_{ii\cdot(i-1)!} = \gamma^2_{iy\cdot(i-1)!} S_{yy\cdot(i-1)!} \tag{10.62}$$

Consider Equation 10.53 for $j = y$ and $r = i$. This gives

$$S_{yy\cdot i!} = S_{yy\cdot(i-1)!} - |L_{iy}|^2 S_{ii\cdot(i-1)!} \tag{10.63}$$

Substitution from Equation 10.62 shows that

$$S_{yy\cdot i!} = S_{yy\cdot(i-1)!}(1 - \gamma^2_{iy\cdot(i-1)!}) \tag{10.64}$$

As special cases,

$$
\begin{aligned}
S_{yy\cdot 1} &= S_{yy}(1 - \gamma^2_{1y}) \\
S_{yy\cdot 2!} &= S_{yy\cdot 1}(1 - \gamma^2_{2y\cdot 1}) = S_{yy}(1 - \gamma^2_{1y})(1 - \gamma^2_{2y\cdot 1}) \\
S_{yy\cdot 3!} &= S_{yy\cdot 2!}(1 - \gamma^2_{3y\cdot 2!}) \\
&= S_{yy}(1 - \gamma^2_{1y})(1 - \gamma^2_{2y\cdot 1})(1 - \gamma^2_{3y\cdot 2!})
\end{aligned}
\tag{10.65}
$$

and so on. Setting $i = q$ proves that the output noise spectrum in the multiple input/output problem of Figure 10.5 or 10.6 satisfies

$$S_{nn} = S_{yy\cdot q!} = S_{yy}(1 - \gamma^2_{1y})(1 - \gamma^2_{2y\cdot 1})\cdots(1 - \gamma^2_{qy\cdot(q-1)!}) \tag{10.66}$$

This formula shows how it is possible to compute S_{nn} from the original given $(q + 1)$ records.

From Equation 10.65 it follows that

$$S_{yy} \geq S_{yy\cdot 1} \geq S_{yy\cdot 2!} \geq \cdots \geq S_{yy\cdot(q-1)!} \geq S_{yy\cdot q!} \tag{10.67}$$

The quantity $S_{yy\cdot r!}$ represents the conditioned (residual) spectrum of S_{yy} which is not due to optimum linear operations from $x_1(t)$ to $x_r(t)$. Stated another way, $S_{yy\cdot r!}$ represents the noise spectrum at the output $y(t)$ in a multiple input/output model when the inputs are $x_1(t)$ to $x_r(t)$. Thus Equation 10.67 verifies the obvious physical condition that the noise spectrum decreases with more inputs.

10.3.4   Multiple Coherence Functions

The multiple coherence function $\gamma^2_{y:x}$ for the multiple input/output problem of Figure 10.5 or 10.6 is defined using a colon notation by

$$\gamma^2_{y:x} = \gamma^2_{y:q!} = \frac{S_{yy} - S_{nn}}{S_{yy}} \tag{10.68}$$

where the subscript x stands for $x_1(t)$ to $x_q(t)$. Substitution from Equation 10.66 proves the general relation

$$\gamma_{y:q!}^2 = 1 - [(1 - \gamma_{1y}^2)(1 - \gamma_{2y \cdot 1}^2) \cdots (1 - \gamma_{qy \cdot (q-1)!}^2)] \qquad (10.69)$$

showing how $\gamma_{y:q!}^2$ is composed of underlying ordinary and partial coherence functions. As special cases,

$$\begin{aligned}
\gamma_{y:1}^2 &= 1 - [1 - \gamma_{1y}^2] = \gamma_{1y}^2 \\
\gamma_{y:2!}^2 &= 1 - [(1 - \gamma_{1y}^2)(1 - \gamma_{2y \cdot 1}^2)] \\
\gamma_{y:3!}^2 &= 1 - [(1 - \gamma_{1y}^2)(1 - \gamma_{2y \cdot 1}^2)(1 - \gamma_{3y \cdot 2!}^2)]
\end{aligned} \qquad (10.70)$$

When the inputs are mutually uncorrelated, Equation 10.69 reduces to the known result

$$\gamma_{y:q!}^2 = \gamma_{1y}^2 + \gamma_{2y}^2 + \cdots + \gamma_{qy}^2 \qquad (10.71)$$

The multiple coherent output spectrum is defined by

$$S_{y:x} = S_{y:q!} = S_{yy} - S_{nn} = \gamma_{y:q!}^2 S_{yy} \qquad (10.72)$$

Here $S_{y:q!}$ represents the coherent output spectrum at $y(t)$ due to the inputs $x_1(t)$ to $x_q(t)$, while the noise output spectrum is given by

$$S_{yy \cdot q!} = S_{nn} = S_{yy}(1 - \gamma_{y:q!}^2) \qquad (10.73)$$

Observe also that $\gamma_{y:q!}^2 = \gamma_{(q+1):q!}^2$ can be extended by replacing $y(t)$ by any of the $x_i(t)$ to obtain $\gamma_{i:(i-1)!}^2$. This quantity $\gamma_{i:(i-1)!}^2$ is the multiple coherence function for the input $x_i(t)$, considered as an output, due to the $(i-1)$ preceding input records $x_1(t)$ to $x_{i-1}(t)$. This is not the same as the multiple coherence function for the input $x_i(t)$, considered as an output, due to all of the other possible $(q-1)$ input records excluding $x_i(t)$. The latter situation would be a $(q-1)$ multiple input/output problem where the output record $y(t)$ is now the particular $x_i(t)$.

The multiple coherence formula proved in Equation 10.69 offers new ways to interpret physically why the multiple coherence function may be low or high at certain frequencies of interest, in terms of associated ordinary and partial coherence functions at these frequencies. Different equivalent formulas can be written down by interchanging the order of the input records that will yield the same multiple coherence function.

In place of Equations 10.66 and 10.69, one can write

$$S_{yy \cdot q!} = S_{yy} \prod_{i=1}^{q} (1 - \gamma_{iy \cdot (i-1)!}^2) \qquad (10.74)$$

$$\gamma_{y:x}^2 = 1 - \prod_{i=1}^{q} (1 - \gamma_{iy \cdot (i-1)!}^2) \qquad (10.75)$$

For all $i = 1, 2, \ldots, q$, one obtains

$$\prod_{i=1}^{q} (1 - \gamma_{iy \cdot (i-1)!}^2) \leq 1 - \gamma_{iy \cdot (i-1)!}^2 \tag{10.76}$$

Hence, using Equation 10.75,

$$1 - \gamma_{y:x}^2 \leq 1 - \gamma_{iy \cdot (i-1)!}^2 \tag{10.77}$$

proving that

$$\gamma_{y:x}^2 \geq \gamma_{iy \cdot (i-1)!}^2 \qquad \text{for all } i = 1, 2, \ldots, q \tag{10.78}$$

In other words, the multiple coherence function $\gamma_{y:x}^2$ must be greater than or equal to *each* of the underlying ordinary and partial coherence functions contained in computing $\gamma_{y:x}^2$, namely, γ_{1y}^2, $\gamma_{2y \cdot 1}^2$, $\gamma_{3y \cdot 2!}^2$ and so on.

In general, arbitrary inequalities can exist at any frequency of interest between the various ordinary and partial coherence functions, that is, γ_{1y}^2 may be greater than, equal to, or smaller than $\gamma_{2y \cdot 1}^2$. Similarly, there is no restriction on $\gamma_{2y \cdot 1}^2$ compared to $\gamma_{3y \cdot 2!}^2$ and so on. Suppose, however, that the following ordering occurs:

$$\gamma_{1y}^2 \geq \gamma_{2y \cdot 1}^2 \geq \gamma_{3y \cdot 2!}^2 \geq \cdots \geq \gamma_{qy \cdot (q-1)!}^2 \tag{10.79}$$

It then follows with the aid of Equation 10.67 that

$$\gamma_{1y}^2 S_{yy} \geq \gamma_{2y \cdot 1}^2 S_{yy \cdot 1} \geq \gamma_{3y \cdot 2!}^2 S_{yy \cdot 2!} \geq \cdots \geq \gamma_{qy \cdot (q-1)!}^2 S_{yy \cdot (q-1)!} \tag{10.80}$$

These terms are the ordered partial coherent output terms in a multiple input/output model as shown in Figure 10.15, thus providing another useful physical interpretation for these results.

10.4   SIMULATION OF SPECTRAL DENSITY MATRICES

Reference 10.1 shows how to simulate, via computer calculated linear systems, a prescribed q by q spectral density matrix $\mathbf{S}_{xx}$ as per Equation 10.81. These special computed systems can also be used to estimate all of the conditioned spectral density functions, partial coherence functions, and multiple coherence functions listed in Section 10.3. The basis for this simulation procedure will now be explained. In practice, all calculations would be carried out by digital approximations to these formulas.

$$\mathbf{S}_{xx} = \begin{bmatrix} S_{11} & S_{12} & \cdots & S_{1q} \\ S_{21} & S_{22} & \cdots & S_{2q} \\ \vdots & \vdots & & \vdots \\ S_{q1} & S_{q2} & \cdots & S_{qq} \end{bmatrix} \tag{10.81}$$

As in Equation 10.8, for stationary (ergodic) random data, the individual terms

$$S_{ij} = S_{x_i x_j} = \frac{1}{T} E[X_i^* X_j] \qquad i, j = 1, 2, \ldots, q \qquad (10.82)$$

where $X_i = X_i(f)$ is the Fourier transform of $x_i = x_i(t)$ defined by Equation 10.4.

Conceptually, it is assumed that $\mathbf{S}_{xx}$ can be produced by passing a set of q mutually independent white noise sources $w_i = w_i(t), i = 1, 2, \ldots, q$ through a set of suitable linear systems $\{A_{ij}\}$ called an A-matrix. It will be shown that these systems are functions of the given $\{S_{ij}\}$ of Equation 10.82. Any particular system $A_{ij} = A_{ij}(f)$ represents the system from input w_i to output x_j with input index i *before* output index j. In terms of Fourier transforms of input and output quantities, Figure 10.17 displays the A-matrix signal generation model for the case of four input records where $W_i = W_i(f)$ are Fourier transforms of the input records.

In matrix form, the A-matrix is a $q \times q$ triangular matrix as follows:

$$\mathbf{A}_{wx} = \begin{bmatrix} A_{11} & A_{12} & A_{13} & \cdots & A_{1q} \\ 0 & A_{22} & A_{23} & \cdots & A_{2q} \\ 0 & 0 & A_{33} & \cdots & A_{3q} \\ \vdots & \vdots & \vdots & & \vdots \\ 0 & 0 & 0 & & A_{qq} \end{bmatrix} \qquad (10.83)$$

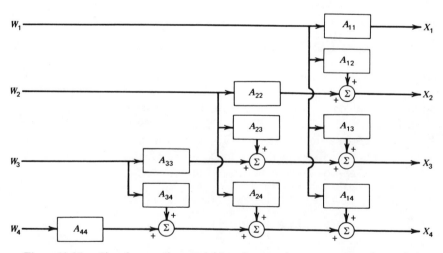

Figure 10.17 Signal generation model (results extend to any number of records.)

All terms on the main diagonal of $\mathbf{A}_{wx}$ are real-valued quantities while the other terms are complex-valued quantities such that

$$A_{ij} = A_{ij}(f) = 0 \qquad \text{for } i > j \tag{10.84}$$

The q mutually independent white noise sources are defined by an identity matrix $\mathbf{S}_{ww} = \mathbf{I}$ where the individual terms

$$S_{w_i w_k} = \frac{1}{T} E[W_i^* W_k] = \begin{cases} 1 & \text{if } i = k \\ 0 & \text{otherwise} \end{cases} \tag{10.85}$$

Thus there are 1's on the main diagonal and 0's everyplace else as follows:

$$\mathbf{S}_{ww} = \begin{bmatrix} 1 & 0 & 0 & \cdots & 0 \\ 0 & 1 & 0 & \cdots & 0 \\ 0 & 0 & 1 & \cdots & 0 \\ \vdots & \vdots & \vdots & & \vdots \\ 0 & 0 & 0 & \cdots & 1 \end{bmatrix} \tag{10.86}$$

To generate such a matrix, the individual Fourier transform input terms can be defined mathematically by

$$W_i(f) = (T)^{1/2} e^{-j\alpha_i(f)} \qquad i = 1, 2, \ldots, q \tag{10.87}$$

to satisfy Equation 10.85. The phase angles $\alpha_i(f)$ are random phase angles for any i and f which are uniformly distributed from 0 to 2π radians.

For q inputs and q outputs, these definitions give the matrix result

$$\begin{bmatrix} X_1 \\ X_2 \\ X_3 \\ \vdots \\ X_q \end{bmatrix} = \begin{bmatrix} A_{11} & 0 & 0 & \cdots & 0 \\ A_{12} & A_{22} & 0 & \cdots & 0 \\ A_{13} & A_{23} & A_{33} & \cdots & 0 \\ \vdots & \vdots & \vdots & & \vdots \\ A_{1q} & A_{2q} & A_{3q} & \cdots & A_{qq} \end{bmatrix} \begin{bmatrix} W_1 \\ W_2 \\ W_3 \\ \vdots \\ W_q \end{bmatrix} \tag{10.88}$$

This is equivalent to the algebraic set of equations

$$X_i = \sum_{k=1}^{i} A_{ki} W_k \qquad i = 1, 2, \ldots, q \tag{10.89}$$

Figure 10.17 is obtained by setting $i = 1, 2, 3, 4$, as follows:

$$\begin{aligned} X_1 &= A_{11} W_1 \\ X_2 &= A_{12} W_1 + A_{22} W_2 \\ X_3 &= A_{13} W_1 + A_{23} W_2 + A_{33} W_3 \\ X_4 &= A_{14} W_1 + A_{24} W_2 + A_{34} W_3 + A_{44} W_4 \end{aligned} \tag{10.90}$$

10.4.1 *Computation of System Terms*

From Equations 10.85 and 10.89, one can prove the following formulas:

$$S_{ij} = \sum_{k=1}^{i} A_{ki}^{*} A_{kj} \qquad i, j = 1, 2, \ldots, q; j \geq i$$

$$S_{ii} = \sum_{k=1}^{i} |A_{ki}|^2 \qquad i = 1, 2, \ldots, q$$

$$(10.91)$$

These relations can be solved for the system terms to yield

$$A_{ii} = [S_{ii \cdot (i-1)!}]^{1/2}$$

$$A_{ij} = \frac{S_{ij \cdot (i-1)!}}{A_{ii}}$$

$$(10.92)$$

Thus, a simple relation exists between the systems $\{A_{ij}\}$, defined by Equation 10.92, and the systems $\{L_{ij}\}$, defined by Equation 10.24, namely

$$A_{ij} = L_{ij} A_{ii} \tag{10.93}$$

This shows that various quantities previously computed in Section 10.3 for solving multiple input/output problems can also be expressed from related terms obtained here for simulating desired spectral density matrices. Similar computational programs can be used for both analysis problems and for simulation problems to provide a close analogy between these two types of engineering applications.

10.4.2 *Computation of Conditioned Quantities*

In general, for any value of $r < i$ with $j \geq i$, conditioned spectral density functions can be computed by

$$S_{ij \cdot r!} = \sum_{k=r+1}^{i} A_{ki}^{*} A_{kj} \tag{10.94}$$

For all i and j with $j \geq i$, partial coherence functions can be computed by

$$\gamma_{ij \cdot (i-1)!}^{2} = \frac{|A_{ij}|^2}{S_{jj \cdot (i-1)!}} \tag{10.95}$$

where

$$S_{jj \cdot (i-1)!} = \sum_{k=i}^{j} |A_{kj}|^2 \tag{10.96}$$

Multiple coherence functions for $j \geq 2$ can be computed by

$$\gamma^2_{j:(j-1)!} = 1 - \frac{|A_{jj}|^2}{S_{jj}} \tag{10.97}$$

where

$$|A_{jj}|^2 = S_{jj\cdot(j-1)!} \tag{10.98}$$

represents the noise spectrum at an output x_j not due to inputs x_1 to x_{j-1}. Note also that

$$S_{jj} = \sum_{i=1}^{j} |A_{ij}|^2 = \sum_{i=1}^{j} |L_{ij}|^2 S_{ii\cdot(i-1)!} \tag{10.99}$$

shows that $|A_{ij}|^2$ represents the uncorrelated outputs in an ordered conditioned multiple input/output model, as per Figure 10.15 when $j = 5$.

REFERENCES

10.1 Dodds, C. J., and Robson, J. D., "Partial Coherence in Multivariate Random Processes," *Journal of Sound and Vibration*, Vol. 42, No. 2, p. 243, 1975.

10.2 Bendat, J. S., "Solutions for the Multiple Input/Output Problem," *Journal of and Vibration*, Vol. 44, No. 3, p. 311, 1976.

10.3 Bendat, J. S., "System Identification from Multiple Input/Output Data," *Journal of Sound and Vibration*, Vol. 49, No. 3, p. 293, 1976.

10.4 Romberg, T. M., "An Algorithm for the Multivariate Spectral Analysis of Linear Systems," *Journal of Sound and Vibration*, Vol. 59, No. 3, p. 395, 1978.

CHAPTER 11

STATISTICAL ERRORS
IN MEASUREMENTS

This chapter deals with statistical errors in the computation of desired quantities from random processes. It is assumed that the data are sample records from stationary (ergodic) or transient processes where the analysis will be done digitally. Results discussed here are for estimates of frequency domain quantities occurring in the analysis of single input/output problems and multiple input/output problems. These include formulas for evaluating autospectral and cross-spectral density functions, ordinary, partial, and multiple coherence functions, coherent output spectra, gain factors, and phase factors of optimum frequency response functions, and other related quantities.

The accuracy of estimates based on sample values can be described by bias errors and random errors as defined in Section 2.4.1 by considering a collection of estimates $\hat{\phi}$ for an unknown true value ϕ. From Equation 2.38, the normalized bias error ε_b and the normalized random error ε_r are given by

$$\varepsilon_b[\hat{\phi}] = \frac{b[\hat{\phi}]}{\phi} \qquad \varepsilon_r[\hat{\phi}] = \frac{\sigma[\hat{\phi}]}{\phi} \tag{11.1}$$

where it is required that ϕ be bounded away from zero. When there is negligible bias error and the random error ε_r is small, from a single estimate, $\hat{\phi}$, there will be approximately 95% confidence that the true value ϕ lies in the interval

$$[\hat{\phi}(1 - 2\varepsilon_r) \leq \phi \leq \hat{\phi}(1 + 2\varepsilon_r)] \tag{11.2}$$

264

This claim can be made even though the actual sampling distribution for $\hat{\phi}$ is theoretically chi-square, F, or some other more complicated distribution, because the sampling distribution for $\hat{\phi}$ when ε_r is small tends to a normal (Gaussian) distribution.

For situations where ε_r is small, it also follows that if one sets

$$\hat{\phi}^2 = \phi^2(1 \pm \varepsilon_r) \tag{11.3}$$

then

$$\hat{\phi} = \phi(1 \pm \varepsilon_r)^{1/2} \approx \phi\left(1 \pm \frac{\varepsilon_r}{2}\right) \tag{11.4}$$

which yields the convenient result that

$$\varepsilon_r[\hat{\phi}^2] \approx 2\varepsilon_r[\hat{\phi}] \tag{11.5}$$

In other words, the random error for squared estimates $\hat{\phi}^2$ is approximately twice the random error for unsquared estimates $\hat{\phi}$. Bias and random errors will now be discussed for spectral density estimates and for more advanced estimates.

To agree with physical applications, the following formulas will be expressed in terms of one-sided spectral density estimates $\hat{G}(f)$ instead of two-sided spectral density estimates $\hat{S}(f)$, as defined in Section 3.2. All normalized bias error and random error formulas in this chapter are exactly the same for one-sided or two-sided estimates.

11.1 SPECTRAL DENSITY FUNCTION ESTIMATES

Consider a sample time history record $x(t)$ from a stationary (ergodic) random process $\{x(t)\}$. An autospectral density function estimate $G_{xx}(f)$ represents a mean square value estimate of $x^2(t)$ in terms of its frequency components lying inside the bandwidth $f - (B_e/2)$ to $f + (B_e/2)$, divided by the bandwidth B_e. This resolution bandwidth B_e should not be confused with the full bandwidth B occupied by the data. Instead, B_e is equivalent to $\Delta f = (1/T)$, the bandwidth resolution used in the digital computations of Section 3.4.2. The autospectral density function estimate

$$\hat{G}_{xx}(f) = \frac{\hat{\psi}_x^2(f, B_e)}{B_e} \tag{11.6}$$

where $\hat{\psi}_x^2(f, B_e)$ is the mean square value estimate of $x(t)$ within the bandwidth B_e centered at frequency f. This mean square value estimate is obtained using a total sample record of length $T_{total} = n_d T$. By digital computations, as described in Equation 3.89, Equation 11.6 is equivalent to

$$\hat{G}_{xx}(f) = \frac{2}{T} E[|X(f, T)|^2] \tag{11.7}$$

where $X(f, T)$ is the finite Fourier transform of $x(t)$ defined for $0 \le t \le T$, and $E[\ ]$ is an ensemble average over n_d different records of length T each. Estimates are obtained at frequencies spaced B_e apart.

11.1.1 *Bias Errors*

From Reference 11.1, as a first order of approximation, the estimate $\hat{G}_{xx}(f)$ will be a *biased* estimate where

$$b[\hat{G}_{xx}(f)] \approx \frac{B_e^2}{24} G''_{xx}(f) \tag{11.8}$$

The quantity $G''_{xx}(f)$ is the second derivative of $G_{xx}(f)$ with respect to f. As such, it will often be quite large in practice at frequencies where sharp peaks occur. The normalized bias error is given by

$$\varepsilon_b[\hat{G}_{xx}(f)] \approx \frac{B_e^2}{24} \left[\frac{G''_{xx}(f)}{G_{xx}(f)} \right] \tag{11.9}$$

Consider a typical single degree-of-freedom system such as a force input-displacement output system described in Equation 1.54. Its frequency response function can be represented by

$$H(f) = \frac{1/k}{1 - (f/f_n)^2 + j2\zeta(f/f_n)} \tag{11.10}$$

where ζ = damping ratio and f_n = undamped natural frequency. If a theoretical white noise input with autospectral density function $G_{ww}(f) = K$, a constant, passes through this system, then the output autospectral density function $G_{xx}(f)$ takes the form

$$G_{xx}(f) = |H(f)|^2 G_{ww}(f) = \frac{K/k^2}{[1 - (f/f_n)^2]^2 + (2\zeta f/f_n)^2} \tag{11.11}$$

This result describes realistic bandwidth-limited white noise data where $G_{ww}(f) = K$ for $0 \le f_1 \le f \le f_2$ with $f_1 \ll f_n \ll f_2$. The peak value of $G_{xx}(f)$ occurs at the resonance frequency f_r and is given by

$$G_{xx}(f_r) = \frac{K}{4k^2\zeta^2(1 - \zeta^2)} \tag{11.12}$$

where

$$f_r = f_n\sqrt{1 - 2\zeta^2} \qquad \text{for } \zeta^2 \leq 0.50 \qquad (11.13)$$

Clearly, if $\zeta^2 \ll 1$, then $f_r \approx f_n$ and

$$G_{xx}(f_r) \approx G_{xx}(f_n) = \frac{K}{4k^2\zeta^2} \qquad (11.14)$$

It will be assumed here for simplicity, as well as because of its importance in applications, that $\zeta^2 \ll 1$ as occurs in cases of highly peaked spectra.

It is now straightforward to determine $G''_{yy}(f)$ at $f = f_r$. This yields the results

$$G''_{xx}(f) \approx \frac{-K}{2k^2\zeta^4 f_r^4} \qquad \frac{G''_{xx}(f_r)}{G_{xx}(f_r)} \approx \frac{-2}{\zeta^2 f_r^2} \qquad (11.15)$$

From Equation 1.59, when $\zeta^2 \ll 1$, the half-power point bandwidth B_r around f_r is given approximately by

$$B_r \approx 2\zeta f_r \qquad (11.16)$$

Hence

$$\frac{G''_{xx}(f_r)}{G_{xx}(f_r)} \approx \frac{-8}{B_r^2} \qquad (11.17)$$

Substituting into Equation 11.9 yields

$$\varepsilon_b[\hat{G}_{xx}(f_r)] \approx \frac{-1}{3}\left(\frac{B_e}{B_r}\right)^2 \qquad (11.18)$$

This simple practical result, which is plotted in Figure 11.1, shows quantitatively how negative bias errors occur in spectral estimates at resonance

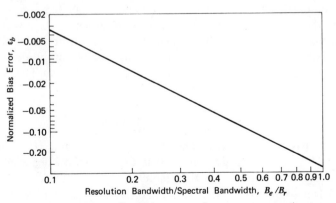

Figure 11.1   Normalized bias error of autospectrum estimates.

frequencies in terms of B_e related to B_r. Such bias errors can usually be ignored if one can choose $B_e = 0.25\, B_r$, where $\varepsilon_b \approx 2.1\%$. A choice of $B_e = 0.50\, B_r$ results in $\varepsilon_b \approx 8.3\%$, which may be marginal. However, if $B_e \approx B_r$, then $\varepsilon_b > 30\%$, which is generally not acceptable.

11.1.2   *Random Errors*

As shown in Reference 11.1, the estimate $\hat{G}_{xx}(f)$ of Equation 11.6 will have a *variance* error

$$\mathrm{Var}[\hat{G}_{xx}(f)] \approx \frac{G_{xx}^2(f)}{B_e\, T_{\text{total}}} \tag{11.19}$$

where $T_{\text{total}} = n_d T$ is the total record length of $x(t)$. This formula is derived under the assumption that $x(t)$ behaves like bandwidth-limited Gaussian white noise over the bandwidth B_e, an assumption that will be closely approximated if B_e is sufficiently small.

An independent derivation of this result is carried out in Reference 11.2 for digital estimates of autospectral density functions as computed by Equation 11.7. The formula is

$$\mathrm{Var}[\hat{G}_{xx}(f)] = \frac{G_{xx}^2(f)}{n_d} \tag{11.20}$$

under the same assumption that $x(t)$ behaves like bandwidth-limited Gaussian white noise over the resolution bandwidth Δf. Since $B_e = \Delta f = (1/T)$ and $T_{\text{total}} = n_d T$, it is clear that $B_e T_{\text{total}} = n_d$. Thus Equations 11.19 and 11.20 are merely two different ways, analog versus digital, to express the same relation.

Reference 11.2 also derives the variance error for cross-spectrum magnitude estimates $|\hat{G}_{xy}(f)|$ computed from n_d pairs of associated sample records $x(t)$ and $y(t)$, where $x(t)$ and $y(t)$ are each of total length $T_{\text{total}} = n_d T$. It is assumed that both $x(t)$ and $y(t)$ behave like bandwidth-limited white noise over the bandwidth $\Delta f = (1/T)$, and that any set of amplitude values of $x(t)$ and $y(t)$ follows a multidimensional Gaussian distribution. The formula is

$$\mathrm{Var}[|\hat{G}_{xy}(f)|] = \frac{|G_{xy}(f)|^2}{\gamma_{xy}^2(f) n_d} \tag{11.21}$$

where $\gamma_{xy}^2(f)$ is the coherence function between $x(t)$ and $y(t)$. Note that Equation 11.21 includes Equation 11.20 as a special case when $x(t) = y(t)$.

Conflicting requirements on the resolution bandwidth $\Delta f = B_e$ can be observed from Equations 11.18 and 11.20. Namely, a small value for Δf is desired to reduce the bias error. However, a large value for Δf is desired to reduce the variance error since n_d will increase as Δf increases when T_{total} is a fixed total length because $n_d = (\Delta f)(T_{total})$.

Equations 11.20 and 11.21 yield the normalized random error formulas

$$\varepsilon_r[\hat{G}_{xx}(f)] = \frac{1}{\sqrt{n_d}} \tag{11.22}$$

$$\varepsilon_r[|\hat{G}_{xy}(f)|] = \frac{1}{|\gamma_{xy}(f)|\sqrt{n_d}} \tag{11.23}$$

where $|\gamma_{xy}(f)|$ is the positive square root of $\gamma_{xy}^2(f)$. It is clear that $\varepsilon_r[\hat{G}_{xx}(f)] \leq \varepsilon_r[|\hat{G}_{xy}(f)|]$ for all f with equality only when $\gamma_{xy}^2(f) = 1$. Equation 11.23 includes Equation 11.22 as a special case when $\gamma_{xy}^2(f) = 1.0$ for all f. The random error formula for the autospectrum estimate $\hat{G}_{xx}(f)$ is independent of frequency, but the random error formula for the cross-spectrum magnitude estimate $|\hat{G}_{xy}(f)|$ depends on frequency. Plots of Equations 11.22 and 11.23 versus the number of averages n_d for various values of $\gamma_{xy}^2(f)$ are presented in Figure 11.2.

Equations 11.18 and 11.22 are the keys to determining the parameter requirements for autospectra estimates. For example, suppose the auto-spectrum of the random vibration of a structure is to be estimated with

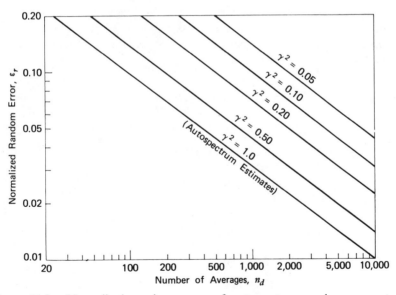

Figure 11.2   Normalized random error of autospectrum and cross-spectrum magnitude estimates.

a maximum normalized bias error of $\varepsilon_b = 0.02$ and a maximum normalized random error of $\varepsilon_r = 0.10$. Assume that it is known that the structure has no resonance frequency below $f_r = 20$ Hz and the damping ratio is about $\zeta = 0.05$. What should be the resolution bandwidth, the number of averages, and the total record length?

For the worst case where $f_r = 20$ Hz, the half-power point bandwidth $B_r \approx 2\zeta f_r = 2$ Hz using $\zeta = 0.05$. Then from Equation 11.18, a resolution bandwidth $B_e = 0.50$ Hz will be suitable to provide a bias error $\varepsilon_b \approx 0.02$. This corresponds to subrecords $T = (1/B_e) = 2$ sec. From Equation 11.22, the requirement that $\varepsilon_r = 0.10$ gives $n_d = 100$. It follows that $T_{total} = n_d T = 200$ sec.

Consider the same structure as above subjected to a broad-band excitation $x(t)$ producing a narrow-band response $y(t)$, where the autospectrum of $y(t)$ has been measured using the same parameters as before. Assume that the expected coherence function $\gamma_{xy}^2(f_r) = 0.70$ at the resonance frequency $f_r = 20$ Hz. What is the required number of averages and total record lengths for $x(t)$ and $y(t)$ to achieve a maximum normalized random error of $\varepsilon_r[|\hat{G}_{xy}(f)|] = 0.10$? What is the expected maximum normalized bias error of $\varepsilon_b[|\hat{G}_{xy}(f)|]$ when the previously specified resolution bandwidth $B_e = 0.50$ Hz is used?

The answer to the second question is the same as before, namely, a resolution bandwidth $B_e = 0.25 B_r$ will produce a normalized bias error of approximately $\varepsilon_b = 0.02$. Answers to the first question follow from Equation 11.23. In particular, $n_d = (100/\gamma_{xy}^2)$ when $\varepsilon_r = 0.10$. Thus $n_d = 143$ when $\gamma_{xy}^2 = 0.70$, as assumed here, and the record lengths for $x(t)$ and $y(t)$ should now be 286 sec instead of 200 sec.

11.2 SINGLE INPUT/OUTPUT PROBLEMS

Consider the simple single input/single output model of Figure 11.3 where

$x(t)$ = measured input signal (assumed noise-free)
$y(t)$ = measured output signal = $v(t) + n(t)$
$v(t)$ = output signal due to $x(t)$ passing through $H_{xy}(f)$
$n(t)$ = output noise
$H_{xy}(f)$ = frequency response function of optimum constant parameter linear system estimating $y(t)$ from $x(t)$

11.2.1 General Considerations

Assume that $x(t)$ and $y(t)$ are the *only* records available for analysis, and that they are representative members of zero-mean stationary random or transient random processes. The assumption that the input $x(t)$ is essentially

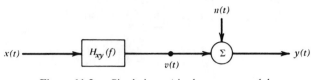

Figure 11.3   Single input/single output model.

noise free is a good assumption in practice when the input can be specified and measured properly by the engineer. However, this is not the case with the output $y(t)$ since the output may include not only the contributions of the particular selected input but also contributions from other known and unknown inputs which were not measured. The noise term $n(t)$ includes these possible effects as well as nonlinear effects and any other deviations which cause $y(t)$ to be different from the $v(t)$ that occurs when $x(t)$ passes through the optimum $H_{xy}(f)$.

For this model, as proved in Section 5.1, the optimum frequency response function estimate is

$$\hat{H}_{xy}(f) = \frac{\hat{G}_{xy}(f)}{\hat{G}_{xx}(f)} \tag{11.24}$$

where $\hat{G}_{xx}(f)$ and $\hat{G}_{xy}(f)$ are estimates of the input autospectral density function and the input/output cross-spectral density function, respectively. The associated ordinary coherence function estimate is calculated by

$$\hat{\gamma}_{xy}^2(f) = \frac{|\hat{G}_{xy}(f)|^2}{\hat{G}_{xx}(f)\hat{G}_{yy}(f)} \tag{11.25}$$

where $\hat{G}_{yy}(f)$ is an estimate of the output autospectral density function.

In computing the estimates $\hat{G}_{xy}(f)$, $\hat{H}_{xy}(f)$, and $\hat{\gamma}_{xy}^2(f)$, it is required to measure $x(t)$ and $y(t)$ using a common time base and to correct for physical propagation time delays τ_1 that may occur between $x(t)$ and $y(t)$ if τ_1 is not negligible compared to the sample record lengths T. Assume

$$x(t) = x(t) \qquad\qquad 0 \le t \le T \tag{11.26}$$

$$y(t) = \begin{cases} \text{arbitrary} & 0 \le t < \tau_1 \\ x(t - \tau_1) & \tau_1 \le t \le T \end{cases} \tag{11.27}$$

Then to a first order of approximation, one can express the cross-correlation estimate $\hat{R}_{xy}(\tau)$ by

$$\hat{R}_{xy}(\tau) \approx \left(1 - \frac{\tau_1}{T}\right) R_{xy}(\tau) \tag{11.28}$$

showing that $\hat{R}_{xy}(\tau)$ is a *biased* estimate of $R_{xy}(\tau)$. It follows that

$$\hat{G}_{xy}(f) \approx \left(1 - \frac{\tau_1}{T}\right)G_{xy}(f) \tag{11.29}$$

$$\hat{H}_{xy}(f) \approx \left(1 - \frac{\tau_1}{T}\right)H_{xy}(f) \tag{11.30}$$

$$\hat{\gamma}_{xy}^2(f) \approx \left(1 - \frac{\tau_1}{T}\right)^2 \gamma_{xy}^2(f) \tag{11.31}$$

The resulting bias error in $\hat{\gamma}_{xy}^2(f)$ for various values of τ_1/T is shown in Table 11.1. To remove these possible bias errors, the signal $y(t)$ should be shifted in time by the delay τ_1 so as to bring $x(t)$ and $y(t)$ into time coincidence. Time delay bias error problems are discussed in Reference 11.3.

The frequency response function can be expressed in polar form as

$$\hat{H}_{xy}(f) = |\hat{H}_{xy}(f)|e^{-j\hat{\phi}_{xy}(f)} \tag{11.32}$$

$|\hat{H}_{xy}(f)|$ = estimate of system gain factor
$\hat{\phi}_{xy}(f)$ = estimate of system phase factor

The quantity $\hat{\phi}_{xy}(f)$ is the phase angle of $\hat{G}_{xy}(f)$ and also the phase that would be assigned to $\hat{\gamma}_{xy}(f)$. For Figure 11.3,

$$\hat{G}_{vv}(f) = |\hat{H}_{xy}(f)|^2\hat{G}_{xx}(f) = \hat{\gamma}_{xy}^2(f)\hat{G}_{yy}(f) \tag{11.33}$$

$$\hat{G}_{nn}(f) = [1 - \hat{\gamma}_{xy}^2(f)]\hat{G}_{yy}(f) \tag{11.34}$$

The estimate $\hat{G}_{vv}(f)$, called the coherent output spectrum, is a computed quantity that is not measured. The measured quantity is

$$\hat{G}_{yy}(f) = \hat{G}_{vv}(f) + \hat{G}_{nn}(f) \tag{11.35}$$

From Equation 5.22, when the optimum $H_{xy}(f)$ is chosen, then $v(t)$ and $n(t)$ will be uncorrelated (uncoherent) so that

$$G_{vn}(f) = E[\hat{G}_{vn}(f)] = 0 \tag{11.36}$$

Table 11.1

Relation of $\hat{\gamma}_{xy}^2$ to (τ_1/T)

τ_1/T	0	0.10	0.20	0.30	0.40	0.50	0.60	0.70
$\hat{\gamma}_{xy}^2/\gamma_{xy}^2$	1.00	0.81	0.64	0.49	0.36	0.25	0.16	0.09

Table 11.2

Relation of $\hat{\gamma}_{xy}^2$ to $[\hat{G}_{vv}/\hat{G}_{nn}]$

$[\hat{G}_{vv}/\hat{G}_{nn}]$	0.11	0.25	0.43	0.67	1.00	1.50	2.33	4.00	9.00
$\hat{\gamma}_{xy}^2$	0.10	0.20	0.30	0.40	0.50	0.60	0.70	0.80	0.90

Here $G_{vn}(f)$ is the true value of the estimate $\hat{G}_{vn}(f)$. Note that the various estimates $\hat{H}_{xy}$, $\hat{\gamma}_{xy}^2$, $\hat{G}_{vv}$, and $\hat{G}_{nn}$ are all found algebraically from the original computed estimates $\hat{G}_{xx}$, $\hat{G}_{yy}$, and $\hat{G}_{xy}$.

Equation 11.33 shows that

$$|\hat{H}_{xy}(f)|^2 = \frac{\hat{G}_{vv}(f)}{\hat{G}_{xx}(f)} = \hat{\gamma}_{xy}^2(f)\frac{\hat{G}_{yy}(f)}{\hat{G}_{xx}(f)} \tag{11.37}$$

Here, from Equation 11.24,

$$|\hat{H}_{xy}(f)| = \frac{|\hat{G}_{xy}(f)|}{\hat{G}_{xx}(f)} \tag{11.38}$$

This is the correct cross-spectra method to estimate the gain factor. An alternative inferior autospectra method to estimate the gain factor is to compute

$$|\hat{H}_{xy}(f)|_a = \left[\frac{\hat{G}_{yy}(f)}{\hat{G}_{xx}(f)}\right]^{1/2} \tag{11.39}$$

Statistical errors in using this autospectra method are discussed in Section 11.3.4.

From Equations 11.33 and 11.34, it follows also that

$$\frac{\hat{G}_{vv}(f)}{\hat{G}_{nn}(f)} = \frac{\hat{\gamma}_{xy}^2(f)}{1 - \hat{\gamma}_{xy}^2(f)} \tag{11.40}$$

$$\hat{\gamma}_{xy}^2(f) = \frac{[\hat{G}_{vv}(f)/\hat{G}_{nn}(f)]}{1 + [\hat{G}_{vv}(f)/\hat{G}_{nn}(f)]} \tag{11.41}$$

Thus $\hat{\gamma}_{xy}^2(f)$ is a function of the output signal-to-noise ratio $[\hat{G}_{vv}(f)/\hat{G}_{nn}(f)]$ as shown in Table 11.2.

11.2.2 Summary of Random Error Formulas

Table 11.3 summarizes practical random error formulas that apply to estimates in single input/output models of Figure 11.3. In Table 11.3, the term $|\gamma_{xy}(f)|$ is the positive square root of $\gamma_{xy}^2(f)$. Note that the random error

Table 11.3

Random Errors in Single Input/Output Problems

Quantity	Random Error, ε_r
$\hat{G}_{xx}(f), \hat{G}_{yy}(f)$	$\dfrac{1}{\sqrt{n_d}}$
$|\hat{G}_{xy}(f)|$	$\dfrac{1}{|\gamma_{xy}(f)|\sqrt{n_d}}$
$\hat{\gamma}_{xy}^2(f)$	$\dfrac{\sqrt{2}[1 - \gamma_{xy}^2(f)]}{|\gamma_{xy}(f)|\sqrt{n_d}}$
$\hat{G}_{vv}(f) = \hat{\gamma}_{xy}^2(f)\hat{G}_{yy}(f)$	$\dfrac{[2 - \gamma_{xy}^2(f)]^{1/2}}{|\gamma_{xy}(f)|\sqrt{n_d}}$
$|\hat{H}_{xy}(f)|$	$\dfrac{[1 - \gamma_{xy}^2(f)]^{1/2}}{|\gamma_{xy}(f)|\sqrt{2n_d}}$

formulas for the basic measurements $\hat{G}_{xx}(f)$ and $\hat{G}_{yy}(f)$ are determined by n_d alone, where n_d represents statistically independent subrecords. For the other advanced measurements, the random error formulas are functions of $\gamma_{xy}^2(f)$ as well as n_d, where results improve as either $\gamma_{xy}^2(f)$ approaches one or n_d becomes large.

To apply these relations to actual data, the unknown true coherence functions $\gamma_{xy}^2(f)$ should be replaced by computed values $\hat{\gamma}_{xy}^2(f)$ of Equation 11.25. This will give appropriate values of ε_r when ε_r is small, say $\varepsilon_r < 0.20$, to substitute into Equation 11.2 to state 95% confidence limits for desired true values.

In the remainder of this chapter the symbol ε_r will be replaced by ε to simplify the notation. The random error formulas in Table 11.3 satisfy the following inequalities:

$$\varepsilon[\hat{G}_{yy}(f)] \le \varepsilon[|\hat{G}_{xy}(f)|] \le \varepsilon[\hat{G}_{vv}(f)] \quad \text{for all } f \qquad (11.42)$$

$$\varepsilon[|\hat{H}_{xy}(f)|] \le \varepsilon[|\hat{G}_{xy}(f)|] \quad \text{for all } f \qquad (11.43)$$

$$\varepsilon[\hat{\gamma}_{xy}^2(f)] \le \varepsilon[|\hat{H}_{xy}(f)|] \quad \text{when } \gamma_{xy}^2(f) \ge (\tfrac{3}{4}) \qquad (11.44)$$

$$\varepsilon[|\hat{H}_{xy}(f)|] \le \varepsilon[\hat{G}_{yy}(f)] \quad \text{when } \gamma_{xy}^2(f) \ge (\tfrac{1}{3}) \qquad (11.45)$$

$$\varepsilon[\hat{\gamma}_{xy}^2(f)] \le \varepsilon[\hat{G}_{yy}(f)] \quad \text{when } \gamma_{xy}^2(f) \ge (\tfrac{1}{2}) \qquad (11.46)$$

Proofs of these formulas and other formulas in this chapter are in References
11.1 and 11.2 or are obvious extensions of these matters.

11.2.3   *Ordinary Coherence Function Estimates*

If n_d distinct averages are used to compute estimates of autospectra and
cross-spectra from the original data, then

$$\varepsilon[\hat{\gamma}^2_{xy}(f)] = \frac{\sqrt{2}[1 - \gamma^2_{xy}(f)]}{|\gamma_{xy}(f)|\sqrt{n_d}} \qquad (11.47)$$

Figure 11.4 plots this equation for various values of $\gamma^2_{xy}(f)$ as a function of n_d.
Table 11.4 lists appropriate values that must be satisfied between $\gamma^2_{xy}(f)$
and n_d to achieve $\varepsilon[\hat{\gamma}^2_{xy}(f)] = 0.10$.

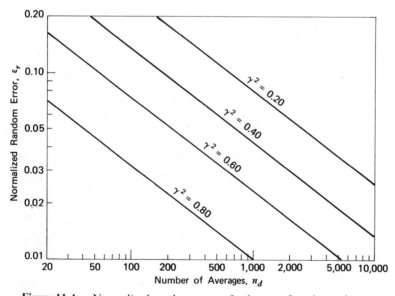

Figure 11.4   Normalized random error of coherence function estimates.

Table 11.4

Conditions for $\varepsilon[\hat{\gamma}^2_{xy}(f)] = 0.10$

$\gamma^2_{xy}(f)$	0.20	0.30	0.40	0.50	0.60	0.70	0.80
n_d	640	327	180	100	54	26	16

These results show for sufficiently large values of γ_{xy}^2 that coherence function estimates $\hat{\gamma}_{xy}^2$ can be more accurate than the autospectra and cross-spectra estimates used in their computation. To be specific, if $\gamma_{xy}^2 = 0.80$ and $n_d = 100$, then $\varepsilon[\hat{G}_{xx}] = \varepsilon[\hat{G}_{yy}] = 0.10$ and $\varepsilon[|\hat{G}_{xy}|] \approx 0.11$, whereas $\varepsilon[\hat{\gamma}_{xy}^2] \approx 0.03$.

11.2.4 Coherent Output Spectrum Estimates

If n_d distinct averages are used to compute estimates of autospectra and cross-spectra from the original data, then

$$\varepsilon[\hat{G}_{vv}(f)] = \frac{[2 - \gamma_{xy}^2(f)]^{1/2}}{|\gamma_{xy}(f)|\sqrt{n_d}} \qquad (11.48)$$

Note that $\varepsilon[\hat{G}_{vv}(f)] \geq \varepsilon[\hat{G}_{yy}(f)] = (1/\sqrt{n_d})$ for all f with equality only when $\gamma_{xy}^2(f) = 1$. Figure 11.5 plots Equation 11.48 for various values of

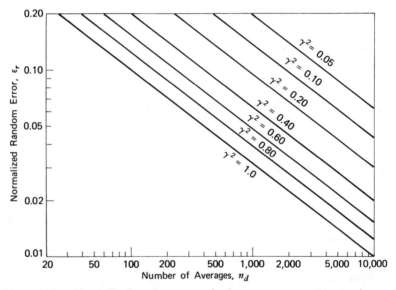

Figure 11.5 Normalized random error of coherent output spectrum estimates.

Table 11.5

Conditions for $\varepsilon[\hat{G}_{vv}(f)] = 0.10$

$\gamma_{xy}^2(f)$	0.20	0.30	0.40	0.50	0.60	0.70	0.80	0.90	1.00
n_d	900	567	400	300	234	186	156	123	100

$\gamma_{xy}^2(f)$ as a function of n_d. Table 11.5 lists appropriate values that must be satisfied between $\gamma_{xy}^2(f)$ and n_d to achieve $\varepsilon[\hat{G}_{vv}(f)] = 0.10$.

11.2.5 Overall Level Estimates

The overall level (output mean square value) as determined from the output spectrum estimate $\hat{G}_{yy}(f)$ is given by

$$\hat{\psi}_y^2 = \int_0^{f_c} \hat{G}_{yy}(f)\, df = \hat{\psi}_v^2 + \hat{\psi}_n^2 \qquad (11.49)$$

where f_c is the Nyquist cutoff frequency, that is, $f_c = (1/2\Delta t) = (N/2)\Delta f$ where $T = N\Delta t$ and $\Delta f = (1/T)$. On the other hand, the overall level of the output due to the coherent output spectrum estimate $\hat{G}_{vv}(f) = \hat{\gamma}_{xy}^2(f)\hat{G}_{yy}(f)$ is given by

$$\hat{\psi}_v^2 = \int_0^{f_c} \hat{G}_{vv}(f)\, df = \int_0^{f_c} \hat{\gamma}_{xy}^2(f)\hat{G}_{yy}(f)\, df \qquad (11.50)$$

Clearly $\hat{\psi}_v^2 \leq \hat{\psi}_y^2$ with equality if and only if $\hat{\gamma}_{xy}^2(f) = 1$ for all frequencies.

As a guideline only, by letting f_c replace $B_e = \Delta f$, the quantity $n_d = B_e T_{\text{total}}$ in Equation 11.20 becomes $n_d(f_c/\Delta f) = n_d(N/2)$ to give

$$\varepsilon[\hat{\psi}_y^2] \approx \frac{\varepsilon[\hat{G}_{yy}(f)]}{\sqrt{N/2}} \approx \frac{\sqrt{2}}{\sqrt{Nn_d}} \qquad (11.51)$$

This result is *not* precise because over the wide frequency range from 0 to f_c, data are not expected to behave like bandwidth-limited Gaussian white noise. Similarly,

$$\varepsilon[\hat{\psi}_v^2] \approx \frac{\text{Ave. } \varepsilon[\hat{G}_{vv}(f)]}{\sqrt{N/2}} \qquad (11.52)$$

where the average is taken with respect to all f in the range $0 \leq f \leq f_c$. For all values of f,

$$\varepsilon[\hat{G}_{vv}(f)] \geq \varepsilon[\hat{G}_{yy}(f)] \qquad (11.53)$$

Hence

$$\varepsilon[\hat{\psi}_v^2] \geq \varepsilon[\hat{\psi}_y^2] \qquad (11.54)$$

To illustrate the use of Equations 11.51 and 11.52, suppose $\varepsilon[\hat{G}_{yy}] = 8\%$, Ave. $\varepsilon[\hat{G}_{vv}(f)] = 16\%$, and $N = 128$. Then $\varepsilon[\hat{\psi}_y^2] \approx 1\%$ and $\varepsilon[\hat{\psi}_v^2] \approx 2\%$. Thus relatively large errors in output spectrum estimates or in coherent output spectrum estimates will have relatively small errors in overall level estimates when the number of data values N is sufficiently large.

11.3 FREQUENCY RESPONSE FUNCTION ESTIMATES

The proper and recommended way to estimate the complete frequency response function (transfer function) for both the gain factor and phase factor is by the cross-spectra method of Equation 11.24.

11.3.1 Bias Errors

For the single input/single output case, the estimation of a frequency response function using Equation 11.24 will generally involve bias errors from a number of sources as follows:

1. Bias due to propagation time delays.
2. Nonlinear and/or time-varying system parameters.
3. Bias in autospectral and cross-spectral density estimates used to compute the frequency response function.
4. Measurement noise at the input point. By Equation 11.24, no bias occurs from uncorrelated noise at the output point.
5. Other inputs that are correlated with the measured input. No bias occurs if the other possible inputs are uncorrelated with the measured input since their effect is to act as uncorrelated noise at the output point.

These various matters are discussed in Sections 4.2, 5.2 and 9.2.

11.3.2 Gain Factor Estimates

If n_d distinct averages are used to compute estimates of autospectra and cross-spectra from the original data, then

$$\varepsilon[|\hat{H}_{xy}(f)|] = \frac{[1 - \gamma_{xy}^2(f)]^{1/2}}{|\gamma_{xy}(f)|\sqrt{2n_d}} \qquad (11.55)$$

Figure 11.6 plots this equation for various values of $\gamma_{xy}^2(f)$ as a function of n_d. Table 11.6 lists appropriate values that must be satisfied between $\gamma_{xy}^2(f)$ and n_d to achieve $\varepsilon[|\hat{H}_{xy}(f)|] = 0.10$. Table 11.6, like Table 11.4, shows for sufficiently large values of γ_{xy}^2 that gain factor estimates can be more accurate than the autospectra and cross-spectra used in their computation. To be specific, if $\gamma_{xy}^2 = 0.80$ and $n_d = 100$, then $\varepsilon[\hat{G}_{xx}] = 0.10$ and $\varepsilon[|\hat{G}_{xy}|] \approx 0.11$. whereas $\varepsilon[|\hat{H}_{xy}|] \approx 0.035$.

From Equation 11.5, the normalized random error for the *squared* gain factor estimate $|\hat{H}_{xy}(f)|^2$ obtained by squaring Equation 11.38 is

$$\varepsilon[|\hat{H}_{xy}(f)|^2] \approx 2\varepsilon[|\hat{H}_{xy}(f)|] \qquad (11.56)$$

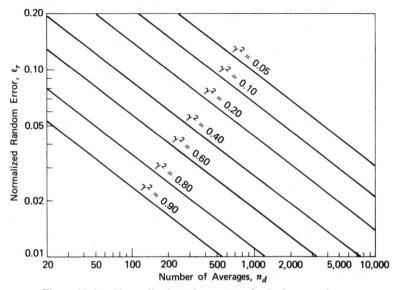

Figure 11.6   Normalized random error of gain factor estimates.

Table 11.6

Conditions for $\varepsilon[|\hat{H}_{xy}(f)|] = 0.10$

$\gamma_{xy}^2(f)$	0.20	0.30	0.40	0.50	0.60	0.70	0.80
n_d	200	117	75	50	34	22	13

Hence using Equation 11.55

$$\varepsilon[|\hat{H}_{xy}(f)|^2] \approx \frac{\sqrt{2}[1 - \gamma_{xy}^2(f)]^{1/2}}{|\gamma_{xy}(f)|\sqrt{n_d}} \tag{11.57}$$

This result is helpful in comparing gain factor estimates obtained by the above cross-spectra method to gain factor estimates obtained by computing only the ratio of autospectra quantities in Section 11.3.4.

11.3.3   *Phase Factor Estimates*

Further results for phase factor estimates $\hat{\phi}_{xy}(f)$ can be obtained by noting that when $\varepsilon[\hat{H}_{xy}(f)]$ is sufficiently small, then the uncertainty $\Delta\hat{\phi}_{xy}(f)$ in radians will satisfy

$$\Delta\hat{\phi}_{xy}(f) \approx \sin \Delta\hat{\phi}_{xy}(f) = \frac{\hat{r}(f)}{|\hat{H}_{xy}(f)|} \approx \varepsilon[|\hat{H}_{xy}(f)|] \tag{11.58}$$

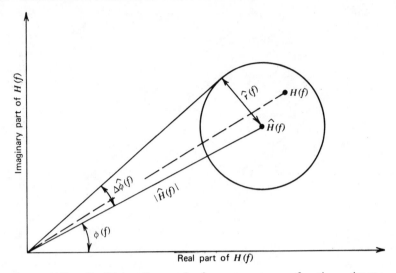

Figure 11.7 Confidence diagram for frequency response function estimates.

The basis for this formula is the confidence diagram of Figure 11.7, where the measured value $\hat{H}$ is shown at the center of the circle with the unknown true value H somewhere inside the circle. From this figure, simultaneous confidence intervals are obtained for both gain factor estimates and phase factor estimates. If the radius $\hat{r}$ of this circle is sufficiently small compared to $|\hat{H}|$, then the 68% confidence intervals are given approximately by

$$\left[\begin{array}{c} |\hat{H}| - \hat{r} \le |H| \le |\hat{H}| + \hat{r} \\ \hat{\phi} - \Delta\hat{\phi} \le \phi \le \hat{\phi} + \Delta\hat{\phi} \end{array} \right] \tag{11.59}$$

and the 95% confidence intervals are given approximately by

$$\left[\begin{array}{c} |\hat{H}| - 2\hat{r} \le |H| \le |\hat{H}| + 2\hat{r} \\ \hat{\phi} - 2\Delta\hat{\phi} \le \phi \le \hat{\phi} + 2\Delta\hat{\phi} \end{array} \right] \tag{11.60}$$

All of these results are functions of frequency. Note from Equation 11.58 that

$$|\hat{H}| \pm \hat{r} = |\hat{H}|(1 \pm \varepsilon[\hat{H}])$$
$$\hat{\phi} \pm \Delta\hat{\phi} = \hat{\phi} \pm \varepsilon[|\hat{H}|] \tag{11.61}$$

In the above equations, the uncertainty $\Delta\hat{\phi}_{xy}(f)$ is measured in radians. A simple conversion to degrees is to use 0.10 radians $\approx 5.7°$. From Equations 11.55 and 11.58,

$$\Delta\hat{\phi}_{xy}(f) \approx \frac{[1 - \gamma_{xy}^2(f)]^{1/2}}{|\gamma_{xy}(f)|\sqrt{2n_d}} \tag{11.62}$$

Table 11.7

Uncertainty $\Delta\hat{\phi}_{xy}(f)$ in Radians for $n_d = 50$

$\gamma_{xy}^2(f)$	0.20	0.30	0.40	0.50	0.60	0.70	0.80
$\Delta\hat{\phi}_{xy}(f)$	0.200	0.153	0.122	0.100	0.082	0.065	0.050

This uncertainty represents the standard deviation of the estimate $\hat{\phi}$. Table 11.7 shows how $\Delta\hat{\phi}_{xy}(f)$ varies as a function of $\gamma_{xy}^2(f)$ when $n_d = 50$.

11.3.4   Gain Factor Estimates by Autospectra Only

If Equation 11.24 is not used to estimate the gain factor, but instead one estimates the *square* of the gain factor by the autospectra ratio of Equation 11.39, then

$$|\hat{H}_{xy}(f)|_a^2 = \frac{\hat{G}_{yy}(f)}{\hat{G}_{xx}(f)} \tag{11.63}$$

Bias errors will now occur due to the nearly always present uncorrelated noise at the output point as pictured in Figure 11.3. This is so serious a problem that Equation 11.63 is *not* recommended in practice unless a nearly ideal linear situation is involved where the coherence function is close to unity at all frequencies of interest.

An analytical representation of the bias error may be developed as follows. From Equations 11.37 and 11.39, to a first order of approximation,

$$E[|\hat{H}_{xy}(f)|_a^2] \approx \frac{|H_{xy}(f)|^2}{\hat{\gamma}_{xy}^2(f)} \tag{11.64}$$

Hence the bias error is given by

$$b[|\hat{H}_{xy}(f)|_a^2] = E[|\hat{H}_{xy}(f)|_a^2] - |H_{xy}(f)|^2$$

$$\approx |H_{xy}(f)|^2 \left[\frac{1 - \hat{\gamma}_{xy}^2(f)}{\hat{\gamma}_{xy}^2(f)} \right] \tag{11.65}$$

and the associated normalized bias error becomes

$$\varepsilon_b[|\hat{H}_{xy}(f)|_a^2] = \frac{b[|\hat{H}_{xy}(f)|_a^2]}{|H_{xy}(f)|^2} \approx \frac{1 - \hat{\gamma}_{xy}^2(f)}{\hat{\gamma}_{xy}^2(f)} \tag{11.66}$$

Similarly, by taking the square root of Equation 11.64,

$$E[|\hat{H}_{xy}(f)|_a] \approx \frac{|H_{xy}(f)|}{|\hat{\gamma}_{xy}(f)|} \tag{11.67}$$

It now follows that the normalized bias error for $|\hat{H}_{xy}(f)|_a$ is given by

$$\varepsilon_b[|\hat{H}_{xy}(f)|_a] \approx \frac{1 - |\hat{\gamma}_{xy}(f)|}{|\hat{\gamma}_{xy}(f)|} \tag{11.68}$$

The bias errors of Equations 11.66 and 11.68 do not occur for gain factor estimates by the cross-spectra method of Equation 11.38. Note that these normalized bias errors are non-negative for all $\hat{\gamma}^2_{xy}(f)$ and that $\hat{\gamma}^2_{xy}(f)$ satisfies Equation 11.41. Numerical results to illustrate these bias errors are listed in Table 11.8 as a function of $\hat{\gamma}^2_{xy}(f)$. Table 11.8 provides quantitative results useful in assessing the bias error in gain factor estimates by autospectra ratios. Clearly, this method should not be used if $\hat{\gamma}^2_{xy}(f) \leq 0.50$, and might be justified with some reservations if $\hat{\gamma}^2_{xy}(f) \geq 0.80$. These normalized bias errors are usually greater than normalized random errors in such measurements, which will be discussed next.

Table 11.8

Normalized Bias Errors in Gain Factor Estimates by Autospectra

$\hat{\gamma}^2_{xy}(f)$	0.50	0.60	0.70	0.80	0.90	0.95
$\varepsilon_b[|\hat{H}_{xy}(f)|^2_a]$	1.00	0.67	0.43	0.25	0.111	0.053
$\varepsilon_b[|\hat{H}_{xy}(f)|_a]$	0.41	0.29	0.20	0.12	0.054	0.026

A derivation similar to that carried out in Reference 11.2 for the random error formulas of Equations 11.55 and 11.57 yields the practical results

$$\varepsilon[|\hat{H}_{xy}(f)|_a] \approx \frac{[1 - \gamma^2_{xy}(f)]^{1/2}}{\gamma^2_{xy}(f)\sqrt{2n_d}} \tag{11.69}$$

$$\varepsilon[|\hat{H}_{xy}(f)|^2_a] \approx \frac{\sqrt{2}[1 - \gamma^2_{xy}(f)]^{1/2}}{\gamma^2_{xy}(f)\sqrt{n_d}} \tag{11.70}$$

These formulas differ from the previous formulas in the replacement of $|\gamma_{xy}(f)|$ by $\gamma^2_{xy}(f)$. Table 11.9 shows appropriate values that must be satisfied between $\gamma^2_{xy}(f)$ and n_d to achieve $\varepsilon[|\hat{H}_{xy}(f)|_a] = 0.10$ using Equation 11.69. Note the significant differences in required n_d from Table 11.6.

From Equations 11.55 and 11.69 one can compare the random error formulas for the gain factor estimate $|\hat{H}_{xy}(f)|$ by the cross-spectra method

Table 11.9

Conditions for $\varepsilon[|\hat{H}_{xy}(f)|_a] = 0.10$

$\gamma_{xy}^2(f)$	0.20	0.30	0.40	0.50	0.60	0.70	0.80
n_d	1000	390	187	100	56	31	16

Table 11.10

Random Error Ratio of Gain Factor Estimates by Autospectra Versus Cross-Spectra Method

$\gamma_{xy}^2(f)$	0.20	0.30	0.40	0.50	0.60	0.70	0.80
$\dfrac{\varepsilon[|\hat{H}_{xy}(f)|_a]}{\varepsilon[|\hat{H}_{xy}(f)|]}$	2.24	1.83	1.58	1.41	1.29	1.20	1.12

to the gain factor estimate $|\hat{H}_{xy}(f)|_a$ by the autospectra method. Independent of the value of n_d

$$\varepsilon[|\hat{H}_{xy}(f)|_a] \approx \frac{\varepsilon[|\hat{H}_{xy}(f)|]}{|\gamma_{xy}(f)|} \qquad (11.71)$$

This result again shows the superiority of the cross-spectra method to the autospectra method in giving a lower random error by the factor $1/|\gamma_{xy}(f)|$. Table 11.10 illustrates Equation 11.71 for various values of $\gamma_{xy}^2(f)$.

11.4 MULTIPLE INPUT/OUTPUT PROBLEMS

Consider the general multiple input/output model of Figure 8.1 or 10.5. All records should be measured simultaneously using a common time base. The first series of steps should be to replace this given model by the conditioned input model of Figure 10.6, as described in Section 10.1. Smooth estimates $\hat{G}_{ij}(f)$ of autospectra and cross-spectra should be obtained from the original given data using n_d averages. All other quantities can then be computed algebraically, as described in Section 10.3. Computation of successive conditioned terms involves subtraction of terms that reduce the number of averages n_d by one each step. Thus multiple coherence function estimates from q inputs will be based on $(n_d - q)$ averages instead of the original n_d averages.

11.4.1   *Multiple Coherence Function Estimates*

For multiple coherence function estimates $\hat{\gamma}^2_{y:x}(f)$ from q inputs, the random error is given by

$$\varepsilon[\hat{\gamma}^2_{y:x}(f)] = \frac{\sqrt{2}[1 - \gamma^2_{y:x}(f)]}{|\gamma_{y:x}(f)|\sqrt{n_d - q}} \qquad (11.72)$$

This is exactly the same form as Equation 11.47 except that the ordinary coherence function $\gamma^2_{xy}(f)$ is replaced by $\gamma^2_{y:x}(f)$ and n_d is replaced by $n_d - q$.

The multiple coherent output spectrum estimate is defined by

$$\hat{G}_{y:x}(f) = \hat{\gamma}^2_{y:x}(f)\hat{G}_{yy}(f) \qquad (11.73)$$

For such estimates, the random error is

$$\varepsilon[\hat{G}_{y:x}(f)] = \frac{[2 - \gamma^2_{y:x}(f)]^{1/2}}{|\gamma_{y:x}(f)|\sqrt{n_d - q}} \qquad (11.74)$$

This is exactly the same form as Equation 11.48 except that $\gamma^2_{y:x}(f)$ replaces $\gamma^2_{xy}(f)$ and $n_d - q$ replaces n_d.

Appropriate guideline formulas for overall level estimates in multiple input/output problems are analogous to Equations 11.49 to 11.54 for the single input/output problem. The main changes are that $(n_d - q)$ replaces n_d, $\gamma^2_{y:x}(f)$ replaces $\gamma^2_{xy}(f)$, and $G_{vv}(f)$ is now due to all of the q multiple inputs instead of only to a single input.

11.4.2   *Partial Coherence Function Estimates*

To simplify the notation, the dependence on f will now be omitted. Figure 10.6 can be broken down into a set of single conditioned input/output models whose spectral nature using one-sided functions is illustrated in Figure 11.8. The top system in Figure 11.8 is the single input/output problem. The succeeding systems are direct extensions where it is obvious which terms should be related to the terms in the top system. Computation of conditioned terms in these lower systems involves subtraction of terms that reduce the number of averages n_d by one each successive step.

For any $i = 1, 2, \ldots, q$, the random error in partial coherence estimates is given by

$$\varepsilon[\hat{\gamma}^2_{iy \cdot (i-1)!}] = \frac{\sqrt{2}[1 - \gamma^2_{iy \cdot (i-1)!}]}{|\gamma_{iy \cdot (i-1)!}|\sqrt{n_d + 1 - i}} \qquad (11.75)$$

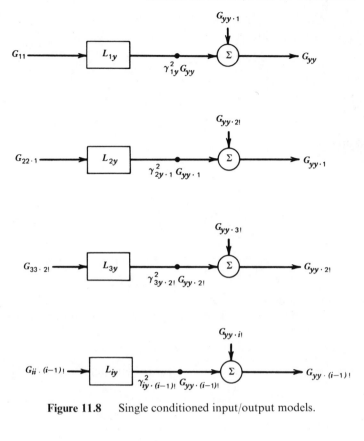

Figure 11.8 Single conditioned input/output models.

As special cases,

$$\varepsilon[\hat{\gamma}_{1y}^2] = \frac{\sqrt{2}[1 - \gamma_{1y}^2]}{|\gamma_{1y}|\sqrt{n_d}}$$

$$\varepsilon[\hat{\gamma}_{2y\cdot 1}^2] = \frac{\sqrt{2}[1 - \gamma_{2y\cdot 1}^2]}{|\gamma_{2y\cdot 1}|\sqrt{n_d - 1}}$$ (11.76)

$$\varepsilon[\hat{\gamma}_{3y\cdot 2!}^2] = \frac{\sqrt{2}[1 - \gamma_{3y\cdot 2!}^2]}{|\gamma_{3y\cdot 2!}|\sqrt{n_d - 2}}$$

Formulas for partial coherent output spectrum estimates also follow analogous to Equation 11.48 for the single input/output problem. These give the normalized random error for the ideal noise-free outputs from the systems L_{iy}, $i = 1, 2, \ldots, q$, in Figure 11.8. To be specific, the partial coherent output spectrum estimates are defined by

$$\hat{G}_{y:i\cdot(i-1)!} = |\hat{L}_{iy}|^2 \hat{G}_{ii\cdot(i-1)!} = \hat{\gamma}^2_{iy\cdot(i-1)!}\hat{G}_{yy\cdot(i-1)!} \tag{11.77}$$

For such estimates, the random error is

$$\varepsilon[\hat{G}_{y:i\cdot(i-1)!}] = \frac{[2 - \gamma^2_{iy\cdot(i-1)!}]^{1/2}}{|\gamma_{iy\cdot(i-1)!}|\sqrt{n_d + 1 - i}} \tag{11.78}$$

As special cases

$$\varepsilon[\hat{G}_{y:1}] = \frac{[2 - \gamma^2_{1y}]^{1/2}}{|\gamma_{1y}|\sqrt{n_d}}$$

$$\varepsilon[\hat{G}_{y:2\cdot1}] = \frac{[2 - \gamma^2_{2y\cdot1}]^{1/2}}{|\gamma_{2y\cdot1}|\sqrt{n_d - 1}} \tag{11.79}$$

$$\varepsilon[\hat{G}_{y:3\cdot2!}] = \frac{[2 - \gamma^2_{3y\cdot2!}]^{1/2}}{|\gamma_{3y\cdot2!}|\sqrt{n_d - 2}}$$

Results from Equations 11.76 and 11.79 are listed in Table 9.2 where the unknown true coherence functions are replaced by computed values.

11.4.3   *Gain and Phase Factor Estimates*

For any $i = 1, 2, \ldots, q$, the random error in gain factor estimates becomes

$$\varepsilon[|\hat{L}_{iy}|] = \frac{[1 - \gamma^2_{iy\cdot(i-1)!}]^{1/2}}{|\gamma_{iy\cdot(i-1)!}|\sqrt{2(n_d + 1 - i)}} \tag{11.80}$$

As special cases,

$$\varepsilon[|\hat{L}_{1y}|] = \frac{[1 - \gamma^2_{1y}]^{1/2}}{|\gamma_{1y}|\sqrt{2n_d}}$$

$$\varepsilon[|\hat{L}_{2y}|] = \frac{[1 - \gamma^2_{2y\cdot1}]^{1/2}}{|\gamma_{2y\cdot1}|\sqrt{2(n_d - 1)}} \tag{11.81}$$

$$\varepsilon[|\hat{L}_{3y}|] = \frac{[1 - \gamma^2_{3y\cdot2!}]^{1/2}}{|\gamma_{3y\cdot2!}|\sqrt{2(n_d - 2)}}$$

When $i = q$, this gives $\varepsilon[|\hat{L}_{qy}|] = \varepsilon[|\hat{H}_{qy}|]$.

If $\varepsilon[\hat{L}_{iy}]$ is sufficiently small, then for any $i = 1, 2, \ldots, q$,

$$\Delta\hat{\phi}_{iy} \approx \varepsilon[|\hat{L}_{iy}|] \tag{11.82}$$

These results are similar to Equation 11.58 and show that $\Delta\hat{\phi}_{iy}$ will be small whenever $\varepsilon[\hat{L}_{iy}]$ is small.

11.4.4 *Special Formulas for Uncorrelated Inputs*

Some special models and formulas will now be discussed that are appropriate for multiple input/output models with uncorrelated inputs and extraneous output noise. To simplify the notation, the dependence on f is again omitted in the quantities shown. Consider the two input/single output model in Figure 11.9. For either X_1 or X_2 *alone*, one can have either of the following separate single input/single output models in Figure 11.10 with extraneous noise. These two single input models *cannot occur simultaneously* since the total output Y is produced by either model alone. Instead, if the first single input model is assumed with X_1 as the input, then the second single input model with X_2 as the input must be as shown in Figure 11.11. Thus the original two input/single output model of Figure 11.9 is equivalent

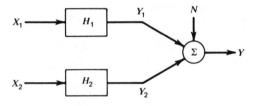

Figure 11.9 Two input/single output model with uncorrelated inputs.

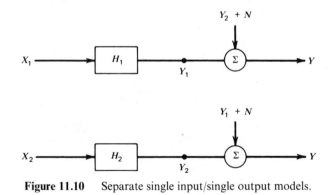

Figure 11.10 Separate single input/single output models.

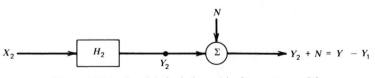

Figure 11.11 Special single input/single output model.

to the following pair of single input/single output models in Figure 11.12, where the order of X_1 or X_2 is arbitrary. The order should be chosen from physical considerations.

Similarly, the general three input/single output model with uncorrelated inputs and extraneous noise is shown in Figure 11.13. This is equivalent to the following combination in Figure 11.14 of three separate single input/single output models, where the order of X_1, X_2, and X_3 is arbitrary. The order again should be chosen from physical considerations. This procedure can be extended to any number of inputs. Equations will now be stated that apply to the general three input/single output model of Figure 11.13. These equations automatically apply to the general two input/single output model of Figure 11.9 when one of the inputs, say X_3, is zero. It should also be clear from this development what the form of the results is for any number of uncorrelated inputs.

For the three input/single output model of Figure 11.13 with uncorrelated inputs X_1, X_2, X_3, and extraneous noise N,

$$Y = Y_1 + Y_2 + Y_3 + N \tag{11.83}$$

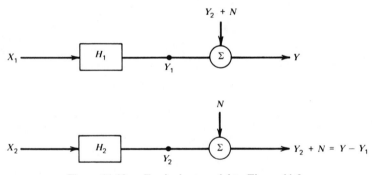

Figure 11.12   Equivalent model to Figure 11.9.

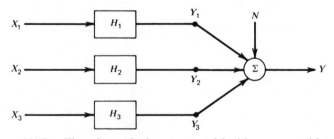

Figure 11.13   Three input/single output model with uncorrected inputs.

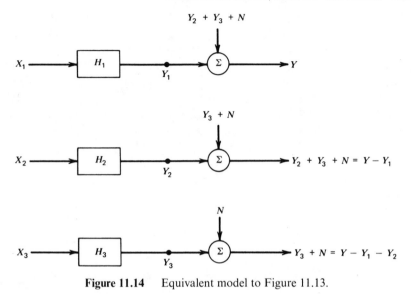

Figure 11.14 Equivalent model to Figure 11.13.

where Y_1, Y_2, Y_3, and N are uncorrelated. These terms satisfy:

$$Y_1 = H_1 X_1 \qquad H_1 = \frac{G_{1y}}{G_{11}} \qquad \gamma_{1y}^2 = \frac{|G_{1y}|^2}{G_{11}G_{yy}}$$

$$Y_2 = H_2 X_2 \qquad H_2 = \frac{G_{2y}}{G_{22}} \qquad \gamma_{2y}^2 = \frac{|G_{2y}|^2}{G_{22}G_{yy}} \qquad (11.84)$$

$$Y_3 = H_3 X_3 \qquad H_3 = \frac{G_{3y}}{G_{33}} \qquad \gamma_{3y}^2 = \frac{|G_{3y}|^2}{G_{33}G_{yy}}$$

The spectral quantities in Figure 11.13 are given by:

$$G_{yy} = G_{y_1y_1} + G_{y_2y_2} + G_{y_3y_3} + G_{nn} \qquad (11.85)$$

$$G_{y_1y_1} = |H_1|^2 G_{11} = \gamma_{1y}^2 G_{yy}$$
$$G_{y_2y_2} = |H_2|^2 G_{22} = \gamma_{2y}^2 G_{yy} \qquad (11.86)$$
$$G_{y_3y_3} = |H_3|^2 G_{33} = \gamma_{3y}^2 G_{yy}$$

For the three separate single input/single output models in Figure 11.14, one requires the following quantities:

$$G_{y_2y_2} + G_{y_3y_3} + G_{nn} = (1 - \gamma_{1y}^2)G_{yy} = G_{yy\cdot 1} \qquad (11.87)$$

$$G_{y_3y_3} + G_{nn} = (1 - \gamma_{1y}^2 - \gamma_{2y}^2)G_{yy} = G_{yy\cdot 2!} \qquad (11.88)$$

$$G_{nn} = (1 - \gamma_{1y}^2 - \gamma_{2y}^2 - \gamma_{3y}^2)G_{yy} = G_{yy\cdot 3!} \qquad (11.89)$$

where

$$G_{yy\cdot 1} = G_{yy} - G_{y_1 y_1}$$
$$G_{yy\cdot 2!} = G_{yy} - G_{y_1 y_1} - G_{y_2 y_2} \tag{11.90}$$
$$G_{yy\cdot 3!} = G_{yy} - G_{y_1 y_1} - G_{y_2 y_2} - G_{y_3 y_3}$$

Since X_1, X_2, and X_3 are uncorrelated, one obtains here:

$$G_{22\cdot 1} = G_{22} \qquad G_{33\cdot 2!} = G_{33}$$
$$G_{2y\cdot 1} = G_{2y} \qquad G_{3y\cdot 2!} = G_{3y} \tag{11.91}$$

Also,

$$\gamma_{2y\cdot 1}^2 = \frac{|G_{2y\cdot 1}|^2}{G_{22\cdot 1} G_{yy\cdot 1}} = \frac{\gamma_{2y}^2}{(1 - \gamma_{1y}^2)} \tag{11.92}$$

$$\gamma_{3y\cdot 2!}^2 = \frac{|G_{3y\cdot 2!}|^2}{G_{33\cdot 2!} G_{yy\cdot 2!}} = \frac{\gamma_{3y}^2}{(1 - \gamma_{1y}^2 - \gamma_{2y}^2)} \tag{11.93}$$

From formulas listed previously, it is now straightforward to write the following random error results for quantities in the three separate single input/single output models in Figure 11.14 that, collectively, are equivalent to the model in Figure 11.13:

$$\varepsilon[\hat{G}_{y_1 y_1}] = \frac{(2 - \gamma_{1y}^2)^{1/2}}{|\gamma_{1y}|\sqrt{n_d}} \tag{11.94}$$

$$\varepsilon[\hat{G}_{y_2 y_2}] = \frac{(2 - \gamma_{2y\cdot 1}^2)^{1/2}}{|\gamma_{2y\cdot 1}|\sqrt{n_d - 1}} = \frac{[2(1 - \gamma_{1y}^2) - \gamma_{2y}^2]^{1/2}}{|\gamma_{2y}|\sqrt{n_d - 1}} \tag{11.95}$$

$$\varepsilon[\hat{G}_{y_3 y_3}] = \frac{(2 - \gamma_{3y\cdot 2!}^2)^{1/2}}{|\gamma_{3y\cdot 2!}|\sqrt{n_d - 2}} = \frac{[2(1 - \gamma_{1y}^2 - \gamma_{2y}^2) - \gamma_{3y}^2]^{1/2}}{|\gamma_{3y}|\sqrt{n_d - 2}} \tag{11.96}$$

$$\varepsilon[|\hat{H}_1|] = \frac{(1 - \gamma_{1y}^2)^{1/2}}{|\gamma_{1y}|\sqrt{2n_d}} \tag{11.97}$$

$$\varepsilon[|\hat{H}_2|] = \frac{(1 - \gamma_{2y}^2)^{1/2}}{|\gamma_{2y}|\sqrt{2n_d}} \tag{11.98}$$

$$\varepsilon[|\hat{H}_3|] = \frac{(1 - \gamma_{3y}^2)^{1/2}}{|\gamma_{3y}|\sqrt{2n_d}} \tag{11.99}$$

REFERENCES

11.1 Bendat, J. S., and Piersol, A. G., *Random Data: Analysis and Measurement Procedures*, Wiley-Interscience, New York, 1971.

11.2 Bendat, J. S., "Statistical Errors in Measurement of Coherence Functions and Input/Output Quantities," *Journal of Sound and Vibration*. Vol. 59, No. 3, p. 405, 1978.

11.3 Seybert, A. F., and Hamilton, J. F., "Time Delay Bias Errors in Estimating Frequency Response and Coherence Functions," *Journal of Sound and Vibration*, Vol. 60, No. 1, p. 1, 1978.

REFERENCES

Abdel-Ghaffer, A. M., and Housner, G. W., "Ambient Vibration Tests of Suspension Bridge," *Journal of the Engineering Mechanics Division*, ASCE, Vol. 104, No. EM 5, p. 983, 1978.

Barnoski, R. L., "Ordinary Coherence Functions and Mechanical Systems," *AIAA Journal of Aircraft*, Vol. 6, No. 4, p. 372, August 1969.

Barrett, S., "On the Use of Coherence Functions to Evaluate Sources of Dynamic Excitation," *Shock and Vibration Bulletin*, No. 49, Part 1, p. 43, 1979.

Begg, R. D., Mackenzie, A. C., Dodds, C. J., and Loland, O., "Structural Integrity Monitoring Using Digital Processing of Vibration Signals," *Proceedings, Offshore Technology Conference*, OTC 2549, Vol. 2, 1976.

Bendat, J. S., and Piersol, A. G., *Random Data: Analysis and Measurement Procedures*, Wiley-Interscience, New York, 1971.

Bendat, J. S., *Principles and Applications of Random Noise Theory*, Wiley, New York, 1958. Reprinted by Krieger Publishing Co., New York, 1977.

Bendat, J. S., "Solutions for the Multiple Input/Output Problem," *Journal of Sound and Vibration*, Vol. 44, No. 3, p. 311, 1976.

Bendat, J. S., "System Identification from Multiple Input/Output Data," *Journal of Sound and Vibration*, Vol. 49, No. 3, p. 293, 1976.

Bendat, J. S., "Statistical Errors in Measurement of Coherence Functions and Input/Output Quantities," *Journal of Sound and Vibration*, Vol. 59, No. 3, p. 405, 1978.

Blake, W. K., and Waterhouse, R. V., "The Use of Cross-Spectral Density Measurements in Partially Reverberant Sound Fields," *Journal of Sound and Vibration*, Vol. 54, No. 4, p. 589, 1977.

Carter, G. C., Knapp, C. H., and Nuttall, A. H., "Estimation of the Magnitude-Squared Coherence via Overlapped Fast Fourier Transform Processing," *IEEE Transactions on Audio and Electroacoustics*, Vol. AU-21, No. 4, p. 337, August 1973.

Carter, G. C., "Time Delay Estimation," NUSC TR-5335, *Naval Underwater Systems Center*, New London, Connecticut, April 1976.

Cooley, J. W., and Tukey, J. W., "An Algorithm for the Machine Calculation of Complex Fourier Series," *Mathematics of Computation*, Vol. 19, No. 90, p. 297, April 1965.

293

Cremer, L., Heckl, M., and Ungar, E. E., *Structure-Borne Noise*, Springer-Verlag, New York, 1973.

Cron, B. J., and Sherman, C. H., "Spatial-Correlation Functions for Various Noise Models," *Journal of the Acoustical Society of America*, Vol. 34, No. 11, p. 1732, November 1962.

Dodds, C. J., and Robson, J. D., "Partial Coherence in Multivariate Random Processes," *Journal of Sound and Vibration*, Vol. 42, No. 2, p. 243, 1975.

Enochson, L. D., "Digital Techniques in Data Analysis," *Noise Control Engineering*, Vol. 9, No. 2, p. 138, November-December 1977.

Halvorsen, W. G., and Bendat, J. S., "Noise Source Identification Using Coherent Output Power Spectra," *Sound and Vibration*, Vol. 9, p. 15, 1975.

Huebner, K. H., *The Finite Element Method for Engineers*, Wiley, New York, 1975.

Lyon, R. H., *Statistical Energy Analysis of Dynamical Systems*, The MIT Press, Cambridge, Massachusetts, 1975.

Marmarelis, P. Z., and Marmarelis, V. Z., *Analysis of Physiological Systems*, Plenum Press, New York, 1978.

Morrow, C. T., "Point-to-Point Correlation of Sound Pressures in Reverberant Chambers," *Shock and Vibration Bulletin*, No. 39, 1969.

Otnes, R. K., and Enochson, L. D., *Applied Time Series Analysis*, Wiley-Interscience, New York, 1978.

Papoulis, A., *Probability, Random Variables, and Stochastic Processes*, McGraw-Hill, New York, 1965.

Papoulis, A., *Signal Analysis*, McGraw-Hill, New York, 1977.

Piersol, A. G., "Use of Coherence and Phase Data Between Two Receivers in Evaluation of Noise Environments," *Journal of Sound and Vibration*, Vol. 56, No. 2, p. 215, 1978.

Rayleigh, J. W. S., *The Theory of Sound*, Vol. 1 Appendix, Dover Publications, New York, 1945.

Romberg, T. M., "An Algorithm for the Multivariate Spectral Analysis of Linear Systems," *Journal of Sound and Vibration*, Vol. 59, No. 3, p. 395, 1978.

Seybert, A. F., and Hamilton, J. F., "Time Delay Bias Errors in Estimating Frequency Response and Coherence Functions," *Journal of Sound and Vibration*, Vol. 60, No. 1, p. 1, 1978.

Stokey, W. F., "Vibration of Systems Having Distributed Mass and Elasticity," Chapter 7, in *Shock and Vibration Handbook*, 2nd Edition, C. M. Harris and C. E. Crede, Eds.), McGraw-Hill, New York, 1976.

Talbot, C. R. S., "Coherence Function Effects on Phase Difference Interpretation," *Journal of Sound and Vibration*, Vol. 39, No. 3, p. 345, 1975.

White, P. H., "Cross-Correlation In Structural Systems: Dispersive and Nondispersive Waves," *Journal of the Acoustical Society of America*, Vol. 45, No. 5, p. 1118, May 1969.

Wilby, J. F., "The Response of Simple Panels to Turbulent Boundary Layer Excitation," AFFDL-TR-67-70, Wright-Patterson Air Force Base, Ohio, 1967.

Yaglom, A. M., *Stationary Random Functions*, Prentice-Hall, New Jersey, 1962.

INDEX

GLOSSARY OF SYMBOLS

b	Bias of [ ]
B	Cyclic frequency bandwidth (Hz)
c	Damping coefficient, propagation velocity
$C_{xx}(\tau)$	Autocovariance function
$C_{xy}(\tau)$	Cross-covariance function
$C_{xy}(f)$	Coincident spectral density function (one-sided)
d	Distance
$E[\;\;]$	Expected value of [ ]
f	Cyclical frequency (Hz)
Δf	Bandwidth resolution (Hz)
$G_{xx}(f)$	Autospectral density function (one-sided)
$G_{xy}(f)$	Cross-spectral density function (one-sided)
$G_{yy \cdot x}(f)$	Conditioned autospectral density function
$G_{x_i y \cdot x_j}(f)$	Conditioned cross-spectral density function
$\mathscr{G}(f)$	Energy spectral density function (one-sided)
$h(\tau)$	Unit impulse response function
$H(f)$	Frequency response function
$\lvert H(f) \rvert$	System gain factor
j	$\sqrt{-1}$, Index
k	Spring constant, index
$L(f)$	Frequency response function for conditioned inputs
m	Mass
n_d	Number of records, number of averages
N	Number of points per record, sample size
$p(x)$	Probability density function
$P(x)$	Probability distribution function
q	Number of inputs
$Q_{xy}(f)$	Quadrature spectral density function (one-sided)
r	Number of outputs
$R_{xx}(\tau)$	Autocorrelation function